How To Make A Noise

a comprehensive guide to synthesizer programming

Simon Cann

Coombe Hill Publishing

Published by Coombe Hill Publishing
33 Melrose Gardens
New Malden
Surrey KT3 3HQ
www.coombehillpublishing.com

ISBN 978-0-9554955-0-2

1.020070717

Table of Contents

Chapter 5
Modulation and Other Ways of Messing with Things 75

Chapter 6
Modulation in Practice **87**

Chapter 7
Frequency Modulation Synthesis **103**

Introduction
Welcome

Thank you for purchasing this book.

How to Make a Noise is a comprehensive guide to programming synthesizers. In particular it focuses on how to program usable sounds that can be used in a musical context.

Some of you may have read the earlier version of this book that was released in early 2005. This new edition has been completely revised and updated: there's more content (including four new refills), more graphics, more sounds, and with the new layout, there are well over twice as many pages. You will also see that it includes a new synthesizer, Surge, which is featured throughout the book. Finally, as should be evident, the book is now available in hard copy format.

The Featured Synthesizers

This book focuses on, and provides sonic examples from, six of the current leading software synthesizers:

- Cameleon 5000 from Camel Audio.

- Rhino from Big Tick

- Surge from Vember Audio

- Vanguard from reFX

- Wusikstation from Wusik dot com, and

- Z3TA+ from Cakewalk

These synthesizers are used to illustrate specific points. Each synth is capable of much more than I am demonstrating in this book. There are other great synthesizers out there—the techniques set out in this book can be applied equally to many of them. However these six have been chosen to illustrate the techniques covered by this book.

If you don't already own all of these synthesizers, I suggest you do. If nothing else, download a demo and try them out. I am also assuming that you will at least be familiar with the manuals for these synths: this book is not a substitute for reading these manuals and it certainly does not attempt to explain how to access any of the synthesizers' features.

The Accompanying Patches

This book relies heavily on examples (there are over 300 patches covered). Theses patches are available from www.noisesculpture.com/htmanpatches at a cost of $10.

You don't need the patches to read and understand the book: in most (but not all) cases the construction of the patches is explained. However, it will save you a lot of hard work if you get them. Even if you don't own any of the synthesizers featured in this book, you can download a demo and use the patches with the demo (although at the time of writing, the patches could not be loaded into the demo of Rhino).

If you purchased the patches that accompany the earlier edition this book, then you can get hold of an updated set of patches (including all of the new patches together with the patches used in the refills) for free. By the time you are reading this book, you should have received an email from me about the update. If you haven't received that email, then something has gone wrong. All you need to do to get the updated patches is request a new download link—you can do this by following the download instructions that came when you purchased the patches. If you have any problems with that, please drop me an email and I'll get everything sorted out.

By the way, if this book is ever revised, then like I have done this time, a new set of patches will be made available to everyone who has purchased the patches, so you can be sure your patches will always be up to date.

Within the book, patches are identified by their name which is Capitalized for clarity. Any suffix or file type indicators (for instance .fxp) are not shown. Surge and Wusikstation patches are individual patches which are not part of a bank and so the patch names are prefaced by a number to keep the correct order.

OK. That's enough introduction: on to the music.

Chapter 1

Getting Started

If you already have some knowledge about how to program synthesizers this chapter may seem over-simplistic. If you haven't got a clue about programming, first make sure you are familiar with the user manual which comes with Vanguard (the synthesizer we will be using in this chapter) and then persevere with this chapter—it will all become clearer.

For now I would like you to be happy tweaking knobs and making sounds even if you don't know what you're doing nor why you're doing it. I will explain the whys and wherefores in the subsequent chapters.

Whether you're an experienced programmer or completely new to the subject, please don't skip this chapter as it raises some of the main themes I want to carry through this book.

In this chapter we are going to build three very simple patches in Vanguard:

■   a bass sound—First Bass

■   a lead sound—Lead Whine, and

■   a "plucky" sound—Arpeggio. You can guess how this is going to be used.

I do not hold these sounds out as being the best you will ever hear—indeed, I am specifically designing them to be quite average (even though Vanguard is capable of making much more interesting sounds). In this chapter I am also limiting myself by basing each of the three sounds on a single sawtooth waveform and will only use a 24 dB low-pass filter.

However, I do want to illustrate two points in particular here:

■   **Programming for a purpose.** If you want/need a sound for a specific piece, you should be able to design a sound to fit rather than hope you can find a preset that is "satisfactory" for the job.

■ **Making the sounds work in context.** You will hear that on their own these three patches sound pretty uninspiring, however, together they work well.

On the former point, I would include tweaking an existing preset as part of the process of designing your own sound. Also, I wouldn't wish to rule out the notion that a preset can provide the correct sound for any track—I just suggest that it may often be easier to create your own sound rather than spend hours looking through your existing banks of presets. With all of these considerations, please do remember that the issue that matters is whether the track you are producing is any good, not how your came about your sound sources.

Synthesizer Architecture

There are many factors in a sound's design, but in essence a sound will usually have three main elements present in its sound (see Figure 1.1):

■ the sound generator (usually an oscillator)

■ the filter (to shape the tone), and

■ the envelope (to control the volume, and perhaps the tone, over time).

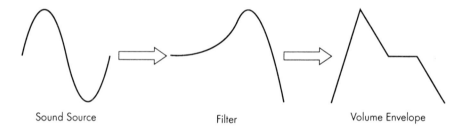

Sound Source   Filter   Volume Envelope

Figure 1.1 The basic signal flow in the sound creation process.

All synthesized sounds emanate from a sound generator of some sort. In the world of software synthesis the sound source is some form of computer code.

Once the sound has been created, it is then "shaped"—both in terms of the tone and the volume over time—to provide a (hopefully) pleasing tone. It is quite possible that the pitch of the sound may be changed over time too. The control sources which effect these changes may include:

■ envelopes, and

■ oscillators (often low frequency oscillators or LFOs).

The term that is applied when one device controls (or changes) another is modulation. So if an LFO controls a filter, perhaps to create a wah-wah type effect, then we would say that the LFO modulates the filter (or more accurately, the LFO modulates the filter's cut-off frequency).

This is just a brief introduction to the workings of synthesizers, we will discuss the various elements in greater detail as we progress.

Start Point for Programming

Many sound designers like to have their own starting point when they begin programming a new sound. This will often be a simple patch set up in a way that they know how everything works.

The patches we are going to make in this chapter are all quite simple, and are based on a saw-tooth wave and a 24 dB low-pass filter which will be my starting point. For the three sounds programmed in this chapter I will describe the construction process starting from my favored Vanguard blank patch, Vanguard Blank. As you can see in Figure 1.2, only a few controls are going to be tweaked to make these patches.

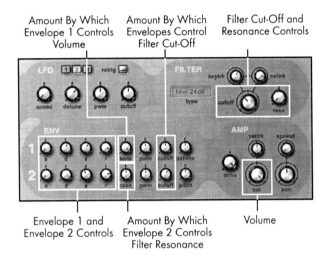

Figure 1.2   All of the sounds in this chapter are controlled with a few knobs.

The patches described in this chapter are included with the patches that accompany this book which are available from www.noisesculpture.com/htmanpatches. If you purchase the patches, you will find them in the Vanguard bank file called Vanguard HTMAN Bank.fxb.

You do not need these patches to follow the examples, but you will find that a lot of the monotony in programming the examples can be avoided if you do purchase the patches. Also, many of the fine details of the more complicated patches are not set out in explicit detail.

For this chapter (and only this chapter), the four patches (and the MIDI file mentioned later) are also available for download for free (from the same location).

First Bass

The intention is that this bass sound will hold a track together but will not dominate. Therefore I am looking for something that is not too aggressive but not too soft. I want the low end but without any bite, so it shouldn't be too bright but neither should it be too dull. I want something really average, but not bland.

Elements of the First Bass Sound

The volume envelope is used to give some shape to the First Bass sound, so the note is more than just on or off. I have set the envelope as follows:

- I am using Envelope One (Env 1) as the volume envelope. The Attack time of the volume envelope is slowed slightly—not by much, but by just enough to take some of the aggression out of the attack. In this case the Attack time (that is the time it takes the sound to go from zero to its maximum) was set to 14ms.

- The Decay time is also slowed down enough that its effect can be heard—here the Decay has little audible effect unless the Sustain level is set considerably below its maximum. In this instance I have set the Sustain level to 42 (where it has a maximum possible value of 127). As a general rule, you will find it much easier to adjust the decay time after you have set the sustain level and if you're unsure about the sustain level, set it too low and then increase it once you have sorted the decay time.

- So now we have set the volume envelope we need to apply it (in other words make it have some effect on the volume). To do this, adjust the Level knob. In this case, I set the Level knob to 127 so that the volume is fully controlled by the envelope.

The filter cut-off was closed down to 963 Hz (by adjusting the Cut-Off knob). This gives quite a dull, lifeless sound so I did two things to improve it. First, the volume envelope was set to slightly modulate (in this context, open up) the filter. For this patch, I set the envelope control of the filter Cut-Off knob to 34 (with a maximum possible value of 127). I also increased the resonance (with the Reso knob) of the filter to 36%—this gives a brighter, slightly less thick quality to the bass note. The effect of the resonance is subtle but it is more noticeable when the envelope modulation is added.

Looking at the modulation of the filter by the volume envelope, the decay time that has been chosen is important:

- if it is very fast you won't hear the effect of the envelope

- if it is quicker than ideal, you'll get a bit of emphasis

- when it is just right (I chose 136ms), you'll hear enough emphasis on the attack of the note (much like hitting a bass guitar string with a plectrum)

- if it is too slow (say over 250ms), you will get a squelchy sort of sound which may be fine in other circumstances but is not what I'm after here. Equally too much filter resonance would have made this sound squelchy—there are several places you need to look if you're trying to dry out "squelch" (for instance, as in this example the decay control or the way the envelope controls the filter—equally, the filter cut-off and resonance control will all have an effect).

Arpeggio

The sound I want for this arpeggio is a thin, bright, staccato (almost percussive) sound. Think of a harpsichord, but not synthetic and not so harsh, and you will be getting close.

Elements of the Arpeggio Sound

For Arpeggio, as I'm going to want the volume envelope (Envelope One) to have a significant effect, I set the Level control in Envelope One to the maximum (127). To get the attack I am after, the Attack in the volume envelope was then set to zero (in other words, the fastest attack).

I set the Sustain level to 27 (out of 127)—any lower and the sustain portion of the patch would not have been audible—thereby making the sound too staccato for my taste in this context. Having set the Sustain level, I set the Decay time. In this instance, I set the Decay time to 357ms—long enough to hear the note, but short enough to give that staccato harpsichord type sound.

The sustain portion of each note is too bright for my taste, so I fixed this with the filter by turning the Filter Cut-Off down to 112 Hz. I could have gone lower (so that the filter would have had even more effect), but then the sound lost some of its clarity when I applied the envelope (see below).

Now we have the right envelope but the sound is quite dull and is virtually silent because the filter is closed. So what I did was to apply the volume envelope to modulate the filter. To do this I set the Cut-Off control in Envelope One to 89 (out of 127). This gives the sound some brightness during its initial attack and decay stages.

While this sound is better, it still wasn't bright enough so I added some resonance in the filter—this makes the sound brighter and thinner. I could have simply turned up the Resonance control in the filter section—this certainly gives a brighter sound. However, because Envelope One (the volume envelope) is modulating the filter it would also mean that the sound gets more reminiscent of a laser gun being shot in a science fiction movie. So I didn't take that option.

Instead, I modulated the filter resonance with the second envelope: as a first step the Reso control in the second envelope bank was turned up to its maximum (127). I then set the envelope—first I set the Attack to zero (the fastest setting). Next I set the Sustain level—here I wanted to ensure there is a bit more brightness on the sustain portion of the note, but not too much, so I set the level at 22 (on a scale of 127). Lastly we needed to set the Decay time—too fast and the effect of the envelope would be lost, too slow and we would get the laser gun effect. 108ms sounded and felt right to me.

The final thing I did with this patch was turn down the volume (the Vol knob in the Amp section) to 90—this balances the levels of the three patches.

Lead Whine

For this lead sound, I was looking for a very muted, almost whining, tone with a characteristic much like a voice through a vocoder or a guitar talk box. I wasn't looking for a burning/searing lead type sound.

Elements of the Lead Whine Sound

As a first step in creating Lead Whine I turned down the Filter Cut-Off to 2.74k Hz and turned up the Resonance to its maximum. If you hold a note and play with the resonance control, as you sweep it from zero to full you will hear a vocal-type sound. I wouldn't go as far as saying one extreme (no resonance) gives an "oooh" type sound and the other extreme (maximum resonance) gives and "aaah" type sound, but we are getting there, slowly. Anyway, leave the Resonance at the maximum.

I wanted a gentle volume envelope (Envelope One), but it must be effective, so I set the Level control in envelope one to the maximum (127). The Attack time was set at its fastest (zero) and I turned the Sustain level down to 29 (out of 127). Finally I adjusted the volume envelope Decay time to 2.82 seconds. This gives a fairly plain sound.

To complete the patch I wanted to capture the character of a guitar talk box—to me this is best represented by the classic sound of morphing vowels. To produce this effect (albeit in a small way) I modulated the Filter Cut-Off frequency with Envelope Two. For a start I set the Attack time of Envelope Two to 923ms and gently applied the envelope to modulate the Cut-Off. To my mind setting the envelope to modulate the Cut-Off at 13 (on a scale of -127 to 127) is just right—any more and the sound becomes too bright and loses that talk box quality.

To complete this sound, I turned down the Sustain level—this means that after the attack and decay portions of envelope two, the filter is less modulated. With the Sustain level set, I tweaked the Decay time—at about 2.63 seconds it felt right to me.

As a final step, the Volume (in the Amp section) was reduced to 65 to stop the output distorting.

Using the three sounds in a mix

In the zip file that contained the four fxp files you will find a MIDI file called Chapter1.mid. Load the MIDI file into your sequencer and load up three instances of Vanguard:

- Assign the first MIDI track (called Lead) to the first instance of Vanguard and load the patch Lead Whine.

- Assign the second MIDI track (called Arpeggio) to the second instance of Vanguard and load the patch Arpeggio.

- Assign the third MIDI track (called Bass) to the third instance of Vanguard and load the patch First Bass).

First play each track solo. You will hear that the sounds are individually not very inspiring. Now play them together and you will hear that the track works well. You are not aware of the static quality of the bass (its sustain phase is masked by the arpeggio). The brightness of the arpeggio is balanced and rounded out by the bass and both of these parts support the lead part.

Why Did We Make These Sounds?

You can hear that these sounds are quite simple and are not that interesting on their own. However, I want to illustrate a few points.

Designed For Purpose

I have created three sounds here for a specific purpose and to work together. I think the end result (in other words the patches used in the context of a track) has been good. If you are creating/editing sounds for a specific purpose you stand a far better chance of getting the right sound which then fits in the mix.

Although these sounds are not interesting on their own, in the context of the track they work well together. While I wouldn't claim that any musical virtuosity is displayed in the MIDI file, the sounds and the track are a perfect marriage. With other sounds—for instance, a searing lead, and a thundering bass—the track probably would not have worked.

Arrangement

Note the arrangement of three parts. You can hear that each sound occupies a different area of the sound spectrum:

- First Bass fills the lower frequencies
- Arpeggio fills the higher frequency range, and
- Lead Whine fills the mid range (which is comparatively uncluttered by the other two parts).

When each sound has its own place in the frequency range, it will be clear and distinct—you may not always want that clarity (for instance, if you are layering), however, you will be more likely to have your mixes criticized for being "muddy" than for being "too sparse".

You can see that I have a certain luxury here—I have full control over both the music and the sound. You may not always have this level of control. However, if you do have this control, then the arrangement of your track will allow you more flexibility—especially if you want to use some "big" sounds that fill a large proportion of the sonic spectrum.

Modern music is highly compressed—if you want your music to stand out, then you will probably compress your music too. If one area of the frequency spectrum is dominant this will mean that the compressors will be disproportionately affected by that part of the spectrum making the rest of the mix comparatively quieter. This is the case even if you're using a multi-band compressor where you will still have one band (or more likely several bands) dominated by one sound. The net effect will be to take energy out of your mix and make it sound comparatively quieter.

Editing Sounds—Balancing Parameters

I see no reason why you shouldn't edit an existing sound to make it suitable for your purposes—there is no law that says all patches have to be created from a blank patch each time. However, I hope these three sounds illustrate that there is generally no one knob that can be tweaked to change a sound. If someone tells you just to change a filter to make a sound less bright, they are either lazy or ignorant—to take the example from these patches, you may be better changing the extent to which the envelope modulates your filter's cut-off. However, you may achieve the effect by just tweaking the filter knob: it all depends on the structure of the patch.

Every patch will be a combination of many factors and you will have to balance several controls to change your patches in a suitable manner. One aspect that these patches particularly highlight is the effect of the decay time on the sounds—this is particularly so for any patch where the envelope controls a filter or a filter's resonance. Play with the decay control in First Bass and Arpeggio and notice the effect it has.

Simplicity

These sounds all use:

- the same filter, and

- the same oscillator

and yet they sound different.

The main differences arise because of the envelope and also the application of the envelope to the filter parameters.

Remember too that these sounds are all monophonic and have no FX.

Programming in Context

None of these sounds is interesting if you listen to it in isolation. None of the sounds works on its own (although the lead might just). There is no doubt that these sounds could be made more interesting, but as I have said, in the context of this track, they work together well.

The three patches are therefore perfect. If I had to suggest improvements, it would be to make the patches more playable—for instance to add some dynamics so that the tone and/or volume change with different levels of velocity.

If you are programming out of context (by which I mean you are programming without your track playing and so you cannot hear all of the other parts and how your patch sounds when it is played and in the context of the mix), then you have a difficult job. You may get a brilliant sound, but I would question whether it will work in the track without further editing.

Let me get to the heart of what I am going to tell you in the rest of the book. In the days before digital audio workstations when hardware ruled the earth, you had to go into a recording studio to make a professional recording. Those days have gone—it is now quite possible for one person to produce a CD on a computer at home which sounds as good as a CD recorded in the most expensive recording studio.

In the days of hardware and recording studios, there were two separate processes: recording and mixing. In today's computer based studios, the processes of writing, recording and mixing have merged. Often there is no longer an engineer on hand to ensure that the frequencies in the mix do not clash, nor a producer to hold your hand through the process.

Instead with computers there are usually one or two people making all of the decisions—without the producer and engineer to guide, this leads to second rate mixes. I have a simple suggestion to address this problem: design your sounds properly and your tracks will mix themselves (in terms of the respective sounds working together). I am not trying to imply that you won't have to do anything with the mix, simply that if you get your sounds right and working within their own areas of the sound spectrum so that they don't fight each other, then it is much easier to balance the respective levels of the elements.

I will return to many of these themes throughout the book.

Chapter 2
Envelopes

I now want to introduce the elements of sound design in a bit more detail. You might be expecting to look at sound sources first—there would be a certain logic given that the sound comes at the start of the signal. However, you're going to have to wait a little while longer, because first we're going to talk about envelopes as these can have a more dramatic effect on a sound than a simple waveform tweak.

Volume Envelopes

The most immediate use for envelopes is controlling volume.

Think of a note played on a piano. When a key is struck, the note goes from silence to the maximum volume instantaneously. From this peak—that is from the moment of impact when the hammer comes into contact with the string—there is an immediate rapid reduction in the volume and then the note reaches a level from which it gradually fades to nothing. A picture of the volume of a piano note over time is set out in Figure 2.1.

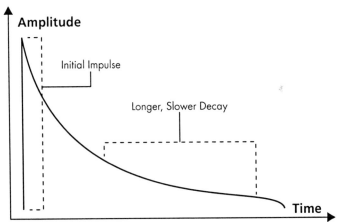

Figure 2.1 The volume envelope of a piano note.

Now if you think about a violin note which slowly fades in, stays at its maximum volume until the note ends, and then gently decays, the volume envelope may look like the image in Figure 2.2.

Figure 2.2
A typical sustained string envelope.

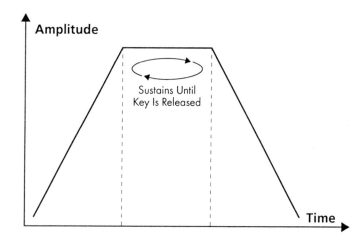

What Else Can Envelopes Do?

Getting more complicated, an envelope changes a level over time—its effect will depend on how it is applied. For instance:

- An envelope could control volume (as we have seen above). Depending on the architecture of the synthesizer, the envelope may control the level of an individual oscillator or the level of a whole patch.

- An envelope may control a filter. If an envelope does control a filter, it will (generally) control the cut-off frequency and so make the sound brighter over time (or, more usually, duller over time, in other words, the sound will start bright and then get duller). If an envelope modulates the filter's resonance then it will control the amount of the resonance—you could set one envelope to close the filter over time and another envelope to increase the amount of resonance while the filter is closing.

- An envelope can also modulate pitch, a common use for this would be to give a short (and subtle) pitch wobble at the start of a note to give the sound more emphasis.

As we will see in Chapter 5: Modulation and Other Ways of Messing with Things and Chapter 6: Modulation in Practice, these are not the only uses for envelopes.

Different synthesizers are designed in different ways. Let's look at some of the more common types of envelope.

ADSR Envelopes

The ADSR envelope is the "classic" envelope (see Figure 2.3). This envelope is used in Surge, Vanguard, and in Wusikstation. There are four main controls:

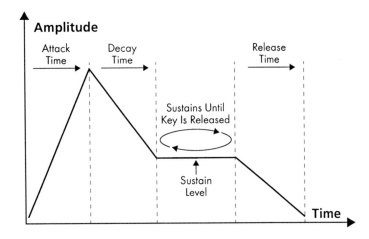

Figure 2.3
The ADSR
envelope.

- **A: Attack time.** This governs the time it takes for the sound to reach its maximum volume after the note is triggered. Using the example of a piano, the attack time would be zero—in other words it would take no time for the note to go from nothing to the maximum volume. For a string type sound the attack may be slower.

- **D: Decay time.** This controls how quickly the sound drops (to the sustain level) after it has reached its maximum volume). Again, using the example of the piano, the decay time would be fast, but it would be longer than the attack time.

- **S: Sustain level.** This is the volume of the sound (or the level of the envelope) while a key is held. In an ADSR envelope, this level stays constant until the key is released—this may be perceived as a weakness if you are using this type of envelope to mimic the behavior of a real instrument where the volume will continue to gently decay over time.

- **R: Release time.** This is the time it takes the sound to decay to zero after a key is released.

You will notice that with this envelope:

- once the sustain part of the envelope has been reached, the note does not decay until the key is released, and

- there is no function in the envelope to determine how long the note sustains when it has reached the sustain level (the only control over this is by releasing the key).

So, assuming you can get the attack and the decay right for a piano type patch, you would not be able to accurately replicate the piano since the envelope does not decay to zero over time—Figure 2.4 illustrates the differences in crude terms.

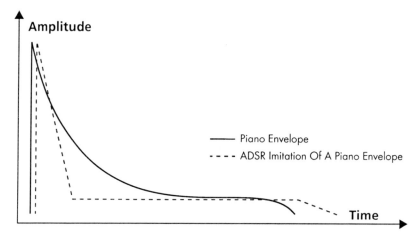

Figure 2.4 The ADSR type envelope may not be the ideal choice for piano type patches.

The next weakness with this type of envelope (and this applies to all envelopes) is that real sounds do not necessarily increase or decrease in a linear manner. Take the example of a slow swelling violin—in practice the attack of the note is likely to have two phases:

■   first the note will go from nothing to a very quiet level very quickly, then

■   the note volume may increase exponentially.

You could almost see the attack as having three phases—a fast phase, followed by a slow phase, followed by another fast phase: Figure 2.5 illustrates this point. It also shows why it may be difficult to use a synthesizer to accurately replicate natural instruments.

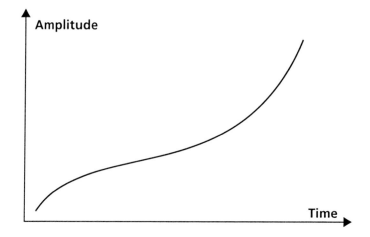

Figure 2.5 The attack phase of a "slow" string sound.

So if the ADSR envelope has limitations, what other choices are there? Lots, but let me just stop you there. I don't want you to think of the ADSR envelope as being limiting—many synthesizers have this style of envelope for good reasons: it works and it is easy to use.

I should also point out at this stage that if you want a piano sound, the optimal solution would be to hire a studio with a piano and a good recording room and get an experienced piano player to record the part. Failing that, there are some excellent sample libraries with highly detailed and playable pianos available, but again I suggest you find a skilled piano player.

DAHDSR Envelopes

Surge has a dedicated filter and volume (amplifier) envelope. These are both conventional ADSR envelopes, although they do have a certain flexibility as the Envelope Curves section, below, will discuss.

However, in Surge's modulation section, there are some more specialist envelopes which can control the modulator, or can be used on their own to modulate any of the modulation destinations. These specialist envelopes have DAHDSR controls (see Figure 2.6):

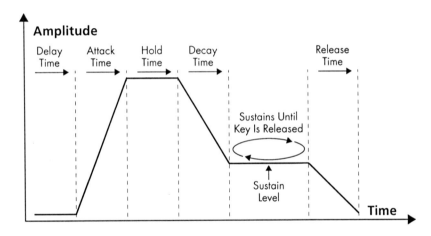

Figure 2.6 The modulation (LFO) envelope in Surge.

- **D: Delay time.** The time before the envelope begins. Generally, this isn't used, but it is useful, especially if you want different elements of a sound to come in after the initial attack.

- **A: Attack time.** This controls the time it takes the note to reach its maximum level once the envelope cycle has begun (in other words, after the delay has ended).

- **H: Hold time.** After the completion of the attack phase, the envelope remains at its maximum level for the Hold time.

- **D: Decay time.** The time it takes (after the Hold phase is complete) for the note to change from the maximum level to the sustain level.

- **S: Sustain level.** This sets the level at which the envelope will remain until the key is released.

- **R: Release time.** The time it takes the note to reach zero after the key is released.

If you set the Delay and Hold controls to zero, these envelopes will work as a conventional ADSR envelope.

One use for this envelope's is to use hold time to increase the time that the sound stays at the level reached after the attack phase has been completed—this can make the sound behave as if the signal is being constrained by a compressor or a limiter. Using this hold stage in moderation can give a sound more punch.

DASSDSR Envelopes

If you look at Figure 2.7, you will see that Z3TA+ takes a different approach to envelope design—at first it may appear complicated, however this just means it is more flexible. This is what the Z3TA+ envelope does:

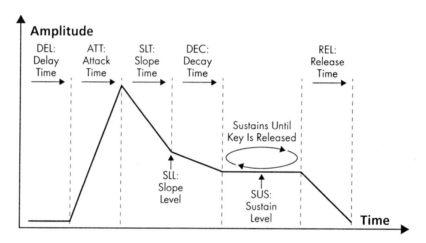

Figure 2.7 The Z3TA+ envelope offers more flexibility than the conventional ADSR envelope.

- **DEL: Delay Time.** This is the time before the envelope begins. Generally, this isn't used, but it is useful, especially if you want different elements of a sound to come in after the initial attack.

- **ATT: Attack Time.** This controls the time it takes the note to reach its maximum level once the envelope cycle has begun (ie after the delay has ended).

- **SLT: Slope Time.** After the completion of the attack phase, the envelope enters the slope stage, this control governs the time it takes the envelope to decay from its maximum level (at the end of the attack phase) to the slope level.

- **SLL: Slope Level.** This sets the level that the envelope will reach at the end of the slope stage.

- **DEC: Decay Time.** This sets the time it takes the note to change from the slope level to the sustain level (note that unlike an ADSR envelope, this change could be an increase or a decrease in level).

- **SUS: Sustain Level.** This sets the level at which the envelope will remain until the key is released.

- **REL: Release Time.** The time it takes the note to reach zero after the key is released.

The Z3TA+ envelope can act like an ADSR envelope (indeed, that may be a good starting point for programming sounds). To do this, set the delay time to zero, the slope time to zero and the slope level to maximum.

Z3TA+ to Imitate a Piano

Z3TA+'s DASSDSR envelope does not provide a perfect envelope for a piano—however, it can make a reasonable imitation as Figure 2.8 shows.

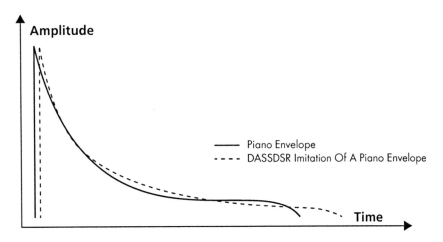

Figure 2.8 The Z3TA+ envelope gets closer to the envelope of a piano.

When imitating a piano (or any other acoustic instrument which decays over time), with the Z3TA+ envelope you can:

- set the attack time, slope time and slope level to resemble the initial impact of the hammer (as I will explain in a moment, you can change the character of the slope), and

- then set a long decay time with a sustain level below the slope level—this will give the character of the note decaying over time.

This still doesn't make a perfect emulation of a piano's envelope (not least since this envelope still does not decay to zero—it will always remain at the sustain level until the key is release). However, it does give the sound designer more flexibility.

Let's build a simple piano type patch in Z3TA+. The emphasis here is on simple: we're not trying to fool any concert pianist. If you have purchased the patches, this patch is Simple Z3TA+ Piano. If you haven't purchased the patches (yet?!?) the settings are listed below.

- First we will load up two oscillators—in the first oscillator load the Piano wave and in the second load a Sine wave. To my mind the Piano wave sounds alright in the lower registers but in the higher registers it sounds a bit too sharp—I use the sine wave to give the tone a bit more roundness (you will find I add pure sine waves to patches quite often). We will look at layering sounds in greater detail in Chapter 4: Sound Sources.

- Next set the Amplitude envelope—this envelope always applies, you do not need to do anything in the modulation matrix to make it have effect:
    - Delay and Attack are both set to zero
    - the Slope Curve is set to exponential (I'll explain this in a moment in the next section, Envelope Curves), the Slope Time to 0.39ms and the Slope Level 51%
    - the Decay time is 0.39ms and the Sustain level 23%, and
    - the Release time is 0.05ms.

- Lastly we will add a bit or reverb and for this
    - engage the Plate reverb
    - set the Size to 45%
    - set Damping to 50%
    - set the High and Low EQ both to -6 dB, and
    - set the Wet/Dry slider to 60%.

And there we have it, a rather crude piano type patch. If you play this it will sound more realistic in the lower registers. The higher registers may be useful if you are trying to play a clavinet type part.

Envelope Curves

In addition to giving control over the envelope, Surge, Wusikstation, and Vanguard all give the sound designer a level of control over the envelopes curves.

Surge and Z3TA+ give three curve options for how a level increases or decreases over time. These curve options are applied:

- to the attack, decay, and release times in Surge, and
- to the attack, slope, decay, and release times in Z3TA+ as Figure 2.9 shows.

The shape of the curves is not specified in Surge (although there is a graphic representation to illustrate what they are up to. However, they are similar to those in Z3TA+ where you can choose:

- linear change, so the level changes uniformly over time
- exponential change, so the change is slow to start but gets more dramatic over time, and

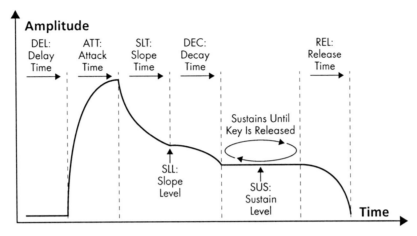

Figure 2.9 The envelope curves in Z3TA+.

- "power" (or logarithmic) change, so the level initially changes quickly, getting slower over time (think of it as being the inverse of the exponential change).

Wusikstation allows different curves to be applied to the attack, decay, and release times in the amplifier envelopes and the modulation envelopes. In any one envelope, all of the curves will be of the same type. In Wusikstation, there are six envelope types to choose from:

- Linear
- Exponential 1×
- Exponential 2×
- Inverted Exponential 1×
- Inverted Exponential 2×
- Inverted Exponential 4×

The different curves give different amounts of steepness to the slopes so allowing for more or less rapid transitions.

Other Envelopes

There are other envelopes which are worth getting to know.

Z3TA+ Pitch Envelope

Z3TA+ also comes with an envelope dedicated to controlling the pitch. The other seven envelopes can control the pitch too, but they are unipolar (that is they only give positive values). The pitch envelope is bipolar—this means it can give positive and negative values in a single cycle as Figure 2.10 shows. In practice this means that the pitch envelope can raise *and* lower a note (where a regular envelope could raise *or* lower a note).

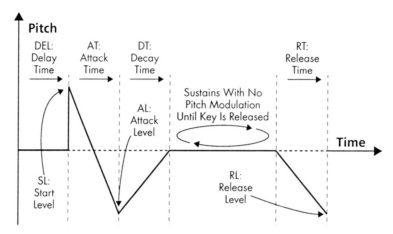

Figure 2.10 The Z3TA+ pitch envelope.

Rhino Envelopes

Rhino takes a different approach and allows you to draw every point on your envelope and to precisely control the shape of the curve between each point. This has two main advantages:

- control—you have very precise control over the design of your envelope, and

- rhythm—you can draw rhythmic envelopes which can be synchronized to the tempo of your track.

The disadvantage is complexity, however, this is largely outweighed by having a graphical interface (see Figure 2.11).

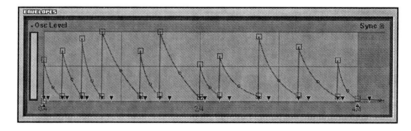

Figure 2.11 Rhino allows you to draw rhythmic envelopes.

Rhino also offers the facility to save envelopes and comes with a bank of envelope presets which can be really useful for creating patches quickly (as some of the examples will demonstrate).

Rhino Piano

Let's create a simple piano patch in Rhino: for those of you with the patches, this is Rhino Piano. The purpose of this patch is twofold—first to introduce Rhino's envelopes and second, to demonstrate how these envelopes can create a more convincing envelope than some of the other options.

To create the sound we will use two oscillators—in oscillator one, we will load the waveform Hard 88 and in oscillator two we will load the waveform FM Tines. Both of these waves can be found under the Electric Piano group of waves. Once loaded, we will drop the pitch of oscillator one by an octave.

The two waveforms are samples of real instruments, so even with a simple envelope (with no volume control) the sound already resembles that of an electric piano. However, if you sustain a note, then it starts to sound unnatural—the volume does not decay as it would in a natural instrument. To remedy this we will load some envelopes.

We could draw some envelopes, but it is much easier and much quicker to load some of Rhino's preset envelopes. In the Envelope Settings > Level folder there is an envelope called Level Piano—I have loaded it for both oscillators. In oscillator one, I then adjusted the curve of the decay (to 25) to give the impression of a slightly faster decay, see Figure 2.12.

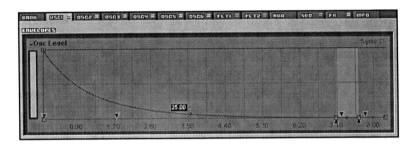

Figure 2.12 The Rhino piano envelope, tweaked to give a sharper decay for oscillator one.

As you can hear (and see), a sample and a preset envelope can be called into action to produce a natural tone quickly and easily.

Cameleon 5000 Envelopes

The Cameleon 5000 takes a slightly different approach to envelopes, but this reflects its different way of creating sound. Cameleon does have a fairly (but not entirely) conventional amplitude envelope and a dedicated modulation envelope, but this is not where the interesting stuff takes place.

Cameleon 5000 is primarily an additive synthesizer, although it does have subtractive synthesis features. With additive synthesis, the sound is made up by a number of sine waves—the combination of sine waves varies over time. Cameleon envelopes have a number of breakpoints: at each breakpoint the combination of sine waves is reconfigured. Between the breakpoints the sine wave configuration changes in a linear manner so that the sound morphs from one combination to the other. A similar process takes place for the noise element of the Cameleon sound.

A fuller explanation of additive synthesis is given in Chapter 9: Additive Synthesis. Figure 2.13 shows the Cameleon envelope.

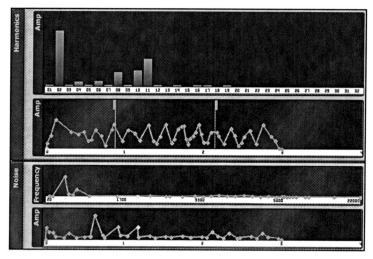

Figure 2.13 Cameleon 5000 envelopes—too difficult to explain in a caption!!

Envelopes and Samples

For sample based synthesizers, such as Wusikstation, the samples will have their own volume envelope. In this case you should think about the interaction between samples and the envelopes. If the wave has a slow attack time, you cannot make it faster simply by choosing a fast attack with your envelope. If you want to get more attack in this situation, you are going to have to change the place that the sample starts to play from—this will have a secondary effect (which may or may not be desirable) of changing the sound and feel of the sample.

Conversely, you can take a sample with a fast attack time and apply a slower envelope. Again, this will change the sound of the sample (which is sort of what we are trying to do...).

For the purpose of this book, when I talk about "sample based" synthesizers, I am referring to machines that can play the whole length of a sample or load a multi-sample (such as Wusikstation or Rhino). I am not calling machines such as Surge and Z3TA+, which have the facility to load single cycle waveform, a sample based synthesizer. Equally I will not refer to Cameleon 5000 as a sample based synthesizer as it re-synthesizes. The distinction between the terms is a fine one and only intended for clarity rather than as qualitative assessments on the strengths of any of the featured synthesizers.

Key Tracking and Envelopes

I want to introduce a new concept here: key tracking (which is also known as pitch tracking). This concept is used in several areas. Essentially key tracking means dynamically changing an element of the sound depending on the pitch. So if we think about the volume envelope of a piano, as the notes get higher they sustain for a shorter period and conversely, as the notes get lower they sustain for longer.

If we want to replicate this behavior on a synthesizer we would use key tracking.

Effect of Envelopes on Sounds

So why did we start by talking about envelopes and not oscillators? Quite simply, because the envelope is one of the most important tools in creating sounds. A lot of nonsense is talked about sound sources—and don't get me wrong, these are important—however, if you use your ears, I think you will find an envelope can change a sound just as much (if not more) than changing an oscillator.

You should also be aware that the volume envelope of a real instrument (unlike a synthesizer with an ADSR envelope) is a ferociously complex thing which can be affected by many factors. No synthesizer envelope will come close to replicating the complexity of the envelope of a real instrument.

Chapter 3
Filters

At its most basic you can think of the filter as being a tone control. A low-pass filter is like the treble control on your stereo—turn it down and the sound gets "duller". However, a filter can do much more for you.

Vanguard gives us one filter—but with lots of options (in fact, probably enough options to suggest that it really gives us two filters in one block). Cameleon gives us two filters, but that's not really the point of the synth. Rhino gives us two filters as does Z3TA+ (although Z3TA+ gives us two stereo filters so that probably makes four), and Surge has six (although we should probably call that 12 if you take account of the stereo option). Oh yeah, Wusikstation gives us 28 filters.

All of the filters are designed to work in slightly different ways: this will affect the sound you hear. One 24 dB low-pass filter will not sound the same as another 24 dB filter, so don't try to make them sound the same—its just too much of a dull job to try.

Filter Types

Let's have a look at the main types of filter that are on offer from our featured synthesizers.

Low-Pass

The low-pass filter (or if you prefer, high-cut filter) allows low frequencies to pass through it, see Figure 3.1.

When a low-pass filter is fully open, all frequencies can pass through it (although some filters do cut the signal even when they are fully open).

As the filter is closed it progressively allows less sound to pass through—you will hear this as the sound becoming duller as the higher frequency elements of the spectrum are filtered out. When the filter is nearly completely closed only the very lowest elements of the frequency spectrum can pass.

27

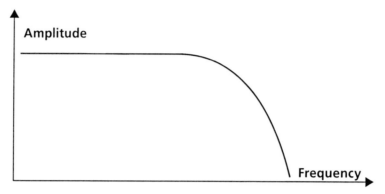

Figure 3.1 A low-pass filter.

The effect that the low-pass filter will have on a sound will vary depending on the source waveform you have selected. If you choose a sine wave, then the effect of the filter will be limited. The sine wave comprises only the fundamental frequency, therefore if the filter cuts this, it cuts the whole sound. However, if you choose a sawtooth wave which has a lot of high frequency information, then the low-pass filter will have a much greater perceived effect on the sound.

However, don't think that if you're using sine waves you won't want to use a filter. For instance, if you are playing chords with a sine wave patch then the individual sine wave notes will interact to produce frequencies beyond the range of the individual sine waves—the filter will affect these frequencies.

High-Pass

The high-pass filter (or low-cut filter) is the reverse of the low-pass filter—it allows high frequencies to pass and progressively cuts out the lower frequencies, see Figure 3.2.

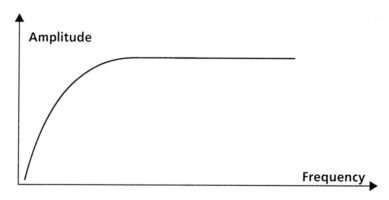

Figure 3.2 A high-pass filter.

As well as sound shaping, high-pass filters have another use: to filter out the junk in the lower end of the mix spectrum. How often have you listened to a track and found it sounds muddy or dull. That could be too much bass. You only get so much dynamic range and without filtering you may be filling your low end needlessly. This means that the key elements—your bass and kick—can't shine.

While high-pass filtering may be noticeable if you play a patch on its own, in a mix any change to the tone of a sound is unlikely to be noticeable except with more extreme cuts. However, the net result of the low end filtering may be to give a cleaner/fuller low end when the bass elements are allowed to come through. You will see that Surge has a high-pass filter hardwired into the signal path to make it easy to clean up your sound in this manner.

Some sound designers also use high-pass filters with resonance (see below) to boost the fundamental tone of a patch. You can do this by tuning the resonance to the fundamental frequency of a note and engaging key tracking so that the resonance boost then follows the played note.

Band-Pass

The band-pass filter acts like a combination of a low-pass filter and a high-pass filter by cutting the frequency spectrum at both ends to only allow a narrow band of sound to pass, see Figure 3.3.

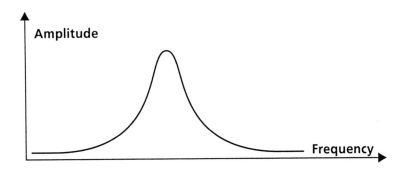

Figure 3.3 A band-pass filter.

The frequency control determines the centre frequency where the full signal is allowed to pass—from that point outwards, the spectrum of sound is progressively cut. At extremes of frequency, the band-pass filter will sound similar to either a high-pass or a low-pass filter.

Band-Pass filters tend to take a lot of energy out of the signal so you may have to boost the level after the filter.

Notch

If the band-pass filter is the equivalent of burning the candle at each end, then the notch filter equates to burning it in the middle. The notch filter cuts the frequencies at its current value, see Figure 3.4.

The notch filter can be used for effect and it can be used surgically in the mix. If you're trying to mix two sounds and they don't sit well together it may be that they're both trying to operate in the same frequency range. In this case, you can "notch out" one of the sounds to allow the other

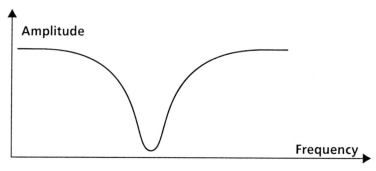

Figure 3.4 A notch filter.

to sit properly—you can do this in your patch design or, as many mixing engineers do, with EQ in the mix.

Formant

Formant filters are usually used to emulate vowel sounds—this is what Z3TA+ does. The Vanguard formant filter has a character that produces more resonant peaks and the Cameleon formant filter is described as being like a powerful multi-band graphic equalizer—it does have the facility to make vocal type sounds, but it can also do more.

Comb Filters

Comb filters work by adding a slightly delayed version of a signal to itself. This causes phase cancellations and can give a slightly "chorused" or metallic type of sound. The spectrum produced by these filters looks like a comb, hence the name.

Of the featured synthesizers, only Rhino and Surge have a comb filter.

Combination Filters

Combination filters are not a different filter type in their own right—instead they are combinations of existing filters. The most obvious example of a combination filter is the notch and low-pass filter in Vanguard.

However, there are other examples that could fall to be considered as combination filters, for instance, Z3TA+'s 24 dB and 36 dB filters are actually stacked 12 dB filters. In normal use this doesn't make a difference, however, by using the separation control, the cut-off frequencies of the stacked filters can be separated (ie one will be raised in relation to the other)—this can give differing resonant peaks (which are discussed in the Resonance section, below) and a different character to the sound.

Filter Parameters

Filters have several parameters and the controls available differ between the six synthesizers featured in this book. However, all of the filters essentially work in the same manner.

Cut-Off Frequency

The cut-off frequency is the point at which the filter starts to have effect, see Figure 3.5. So if you have a low-pass filter and set the cut-off frequency to 8kHz, the sound spectrum above 8kHz will be progressively reduced. However, if you are using a high-pass filter, sounds below the cut-off frequency (8kHz) will be reduced.

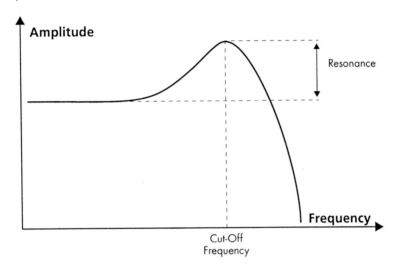

Figure 3.5 A low-pass filter with a resonant peak.

Resonance

Resonance adds some bite to a filter. It works by boosting the sound spectrum around the filter cut-off frequency as Figure 3.5 illustrated. Used in moderation, the effect is subtle and can make a sound appear brighter and/or slightly thinner (or less fat, if you prefer). When used to the extreme the effect is noticeable—most dance records use filter sweeps with high levels of resonance.

At very high resonance settings, some of the filters can exhibit quite extreme behavior—if you're looking for an example, turn on the resonance boost in Z3TA+ and push the resonance right up. Make sure you turn down the output before you try this or you are likely to burn your ears off.

Wusikstation and Z3TA+ have limiters on their filters because of the extreme nature of sounds that can be produced at high filter resonance settings.

Filter Slopes

A low-pass filter progressively reduces the volume of a sound above the cut-off point. The rate at which the sound wave is reduced above the cut-off frequency is determined by the slope of the filter.

If you have a 6 dB/octave filter, it will reduce the level of the sound source by 6 dB at one octave above the cut-off point, 12 dB at two octaves above the cut-off frequency, 18 dB at three octaves above the cut-off.

31

The effect of a 6 dB filter on a sound wave is quite subtle.

A 12 dB/octave filter (sometimes called a 2-pole filter) will reduce the level of the sound wave's volume by 12 dB for each octave for each octave above the cut-off point. This type of filter was often used in some of the Japanese synthesizers from the 1980s.

A 24 dB/octave filter (sometimes called a 4-pole filter) will reduce the level of the sound wave's volume by 24 dB for each octave above the cut-off point. This type of filter was often used in some of the classic US analog synthesizers.

You will see some different filter slopes in Figure 3.6. This is an image drawn by hand and so is not to scale, however, you can get a good idea of the different filter slopes.

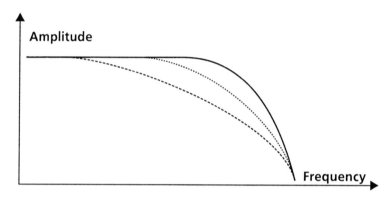

Figure 3.6 Different filter curves—the flatter curves may be more gentle, but in these cases, the filter affects a large portion of the sonic spectrum.

So which is better, a 6 dB, 12 dB, 24 dB or 36 dB filter? That all depends. Take a comparison between a 12 dB and 24 dB filter—you need to think of the context in which the sound will be used. On the face of it, the 24 dB filter might get faster results. However, remember that a 12 dB filter has to work twice as hard and will affect a much greater proportion of the signal to achieve the same reduction that a 24 dB filter will make at any given frequency, so the 12 dB filter may be more appropriate.

There are no hard and fast rules, but you may find that, for instance, a 24 dB filter could be a better starting point when designing bass sounds, whereas a 12 dB filter produces better results with pads or sample based sounds. Then again, you're going have your own considerations and will need to decide whether you're trying to shape the sound or whether you're getting surgical, so you may find another combination that works for you.

Different Synthesizers—Different Filters

Different filter designs operate in different ways and give different results—it is important that you have some idea of what a filter is trying to do to the signal. For instance, some filters (not shown here) will allow you some control over the bandwidth of the resonance—this is quite a fine change, but definitely has an affect on the sound's character.

If you get two synthesizers each with a 12 dB/octave filter, they won't necessarily sound the same. They might initially, but if you start apply envelopes and other modulation sources to the filters, they are likely to work in different ways. There are many reasons for this difference, and while the minor differences between each component may appear comparatively trivial when compared on a side-by-side basis, the combinations—in other words, the final sound—will be markedly different giving each synthesizer its own character.

Think of filter labels a bit like the badges on the back of a car. "2.0" on the back of a car may indicate a two liter engine. Two cars with "2 liter" engines may in practice have engine capacities of 1998 cc and 2021 cc—it's all a matter of design and branding.

You wouldn't expect two cars from different manufacturers with a 2.0 badge on the back to have the same performance. However, it wouldn't be unreasonable to expect that if you have two cars in the same manufacturer's range and one has a 2.0 badge and the other a 3.0 that the bigger engine will accelerate more quickly. But, you would not necessarily expect it to accelerate 50% faster.

Use the same logic with filters: the numbers are labels which don't necessarily translate between different synths—use them as an indication of the differences within the individual synthesizer. At the end of the day, if you want to know how the filters sound and feel, you're just going to have to take the synth out for a test drive.

In practice you will find filters can be arranged in parallel (for instance Rhino) or in series (Surge, Z3TA+ and Wusikstation all give the option to arrange filters in parallel or series). When filters are run in series, the output of each filter is fed into the next filter giving a cumulative effect: two 12 dB filters in series gives the effect of a 24 dB filter.

However, remember that this book is about making music, not mathematics. By design, different filter slopes will achieve their results in different ways. Some will work in a pure linear, mathematical way. Some will cut the sound more sharply after the cut-off point. Some will cut less at the cut-off point, but will cut more (in an inverse exponential curve) as the sound gets higher above the cut-off point. Others will try to emulate classic gear—and some will claim to emulate classic gear knowing that virtually no one has the gear to make an A/B comparison.

Perhaps the best example of differing filter characteristics comes with Surge. Not only does it offer a wide range of filters, within that range, it also offers filters with different characteristics. These different characteristics provide a wide range of sound shaping tools for different sound design tasks. For instance, the main 12 dB/octave and 24 dB/octave low-pass and high-pass filters, and the band-pass filter all come in three flavors:

- Flavor one is a fairly clean filter which can have a strong resonant peak.

- Flavor two is a slightly less clean filter which has a less pronounced resonant peak. This could be considered to have slightly less of a digital sound, but I think it is easier just to think of this as offering a different color.

■ Flavor three is a very smooth filter with a much less prominent resonance. This filter is more suited to subtly shaping the tone and doesn't have the aggressive, dominant character that the other two filters can display when pushed hard.

Filter Configuration and Signal Routing Options

Another factor that has a significant influence on the sound is the routing options. These are perhaps best demonstrated in Surge, although Z3TA+ does also offer several routing options.

Surge also has another feature that the other synthesizers do not offer: feedback. This allows some of the output signal to be fed back into the audio. This is a trick that was often used in the days of modular synthesizers and can help to dirty up or thicken up a signal. It can also give wild results and so should be used with caution.

In Surge you can determine how the signal passes through the filter stage: this gives a huge number of sonic options. In essence, there are three units in Surge's filter stage:

■ filter one

■ filter two

■ a waveshaper (I will discuss this unit further in Chapter 4: Sound Sources)

In addition to this, there is:

■ a feedback loop, and

■ a volume control section with a gain setting and a dedicated controlling envelope.

The gain stage in the filter block has an effect on the feedback loop as well as the output level.

The filters can then be arranged in series or in parallel, as well as in mono or stereo, to give various sonic options. These are the options and some of the reasons why you might consider using those options.

For the sake of completeness, I should also point out that there is a high-pass (low-cut) filter immediately before the output stage to allow you to remove some of the bottom end mud. As this applies to each filter routing configuration, I won't mention it again in this section.

Serial Option One

With serial option one (see Figure 3.7), the input signal flows into filter one, then into the waveshaper, next to filter two, and finally to the amplifier (gain) stage.

This is perhaps the most simple filter configuration. It allows you to filter the signal, apply some waveshaping, and then filter the signal that has been shaped by the waveshaper.

Serial Option Two

With serial option two (see Figure 3.8), the signal flow is exactly the same as in serial option one, however at the end of the chain, the audio signal is then fed-back into the start of the chain. The amount of feedback is controlled by the Feedback slider and the volume/gain set in the amplifier section at the end of the chain.

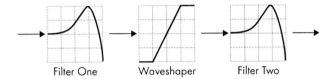

Figure 3.7 Serial option one arrangement of the filter section in Surge.

Like serial option one, this is a straightforward configuration, but this also allows some feedback to be added to dirty up the sound a bit. Positive or negative polarity can be selected which will have differing effects on the resulting sound.

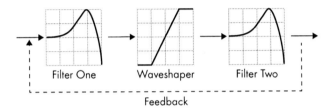

Figure 3.8 Serial option two arrangement of the filter section in Surge. This is the first configuration to allow the signal to be fed back into the audio path.

Serial Option Three

In serial option three (see Figure 3.9), the input signal flows into filter one, then into the waveshaper, and finally to the amplifier (gain) stage. This signal is output, but is also fed back. The feedback loop first passes through filter two before it is fed into the start of the audio chain.

Figure 3.9 Serial option three arrangement of the filter section in Surge. With this filter configuration, the tone of the feedback loop can be shaped with filter two.

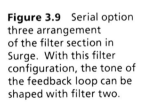 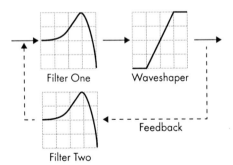

This option allows the tone of the feedback loop to be shaped, giving more control over the feedback. This additional control means that greater amounts of feedback can be used.

Dual Option One

With dual option one (see Figure 3.10), filter one and filter two are arranged in parallel. The signal from each oscillator can be routed to either filter individually or to both filters. Once the signal has passed through the filters, it is mixed and enters the waveshaper. Having been processed in the waveshaper the signal is sent to the amplifier section from where the signal is fed back into the start of the audio chain.

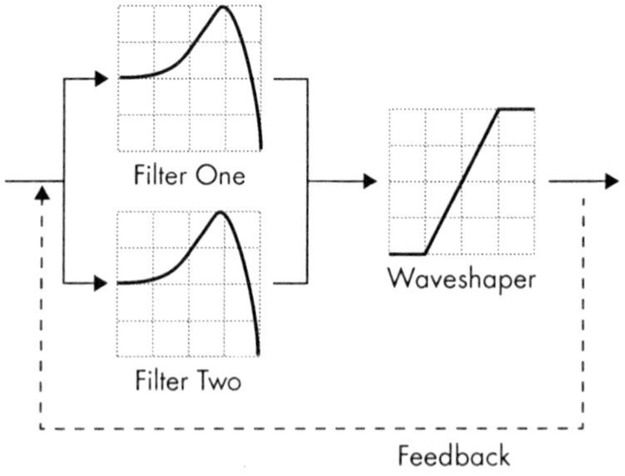

Figure 3.10 Dual option one arrangement of the filter section in Surge. This allows for two separate sounds to be created and processed through different filters.

This option allows for sounds to be processed separately and so gives more layering options.

Dual Option Two

Dual option two (see Figure 3.11), is very similar to dual option one, except that the waveshaper is only applied to filter one and cannot be applied to filter two. So this is the option to choose if you do want to waveshape the output from filter one, while also applying filtering in filter two without applying any shaping.

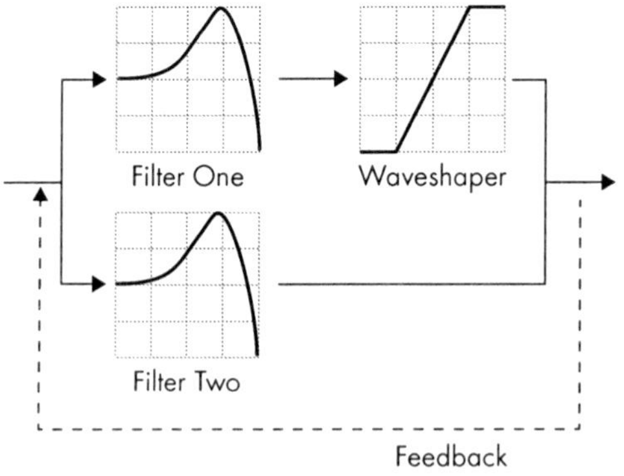

Figure 3.11 Dual option two arrangement of the filter section in Surge. This restricts the application of the waveshaper to the output from filter one only.

Stereo Option

The stereo option (see Figure 3.12), which is labelled L-R sends the left audio channel through filter one and the right channel through filter two. The channels are then kept separate and are passed through two waveshapers (one for each channel) and through two amplifiers (one for each channel). A mixed signal is then fed back.

Figure 3.12 Ths stereo filter configuration in Surge allows the left and right channels to be processed separately.

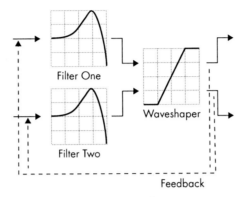

One additional feature that is added for this stereo option (and the wide option, discussed below) is a Width control (which is located in the Output section). By default the left signal is panned hard left and the right signal is panned hard right. This control allows the signal width to be narrow by moving the panning towards the middle of the audio spectrum (or for the panning to be reversed).

Ring Modulation Option

The ring modulation option has the same configuration as dual option one. However, instead of the signals being mixed, they are ring modulated. Ring modulation is discussed further in Chapter 4: Sound Sources.

Wide Option

The wide option has the same configuration as serial option two. However, instead of using a mono signal path, a stereo signal path is used (in other words, the filters and the waveshaper each work as two separate units applied individually to the left and right channels).

In Z3TA+ you can arrange the filters in parallel or in series and you can also route the signal through the Distortion FX unit and back into the filters. While Z3TA+ may not be quite as flexible as Surge in some aspects, it does still give very many options for sound shaping through the choice of different routing options.

Using Filters in Practice

One of the key issues for making a filter sound realistic is the modulation that is applied to it. This is discussed in greater detail in Chapters 5 and 6. A filter which is not modulated will be, and will usually sound, static. The human ear is attuned to natural instruments which have a constantly shifting character and will expect similar characteristics in synthesized sound.

Controlling Filter Cut-Off

When using a filter to shape tone, the main (but not the only) modulation controls are:

- envelopes—these provide an ideal control source to shift the filter over time, and

- velocity—this allows a filter to mimic the properties of natural instruments so that the sound gets brighter with more forceful playing.

You will find that velocity and envelopes are often used in combination when controlling filters.

Filter as a Volume Control

The filter also acts as a volume control—the more a filter closes, the more it attenuates the amplitude of the sound. You can use this to your advantage by setting an envelope to completely close a filter. If we return to our example of a piano note—the note should have a finite length, but an ADSR envelope cannot mimic this behavior. However, a filter controlled by another ADSR envelope could cut the volume completely. Therefore, the combination of two envelopes, one controlling the volume and the other controlling the tone could create a more accurate representation of an acoustic sound.

Key Tracking and Filters

You may want to use key tracking when setting the filter to mimic what happens naturally. If you are designing a patch to resemble the behavior of an acoustic instrument, a single cut-off point may be unnatural: the high notes would be too dull and the low notes to bright. To remedy this you could use key tracking with the filter: this would open the filter at higher frequencies and close it at lower frequencies giving a more natural response. If you want you can have an unnatural response where the filter closes *more* at higher pitches.

Controlling Resonance

Resonance is generally used in four main ways:

- to emphasize the filter cut-off, and in particular, to emphasize a changing filter-cut-off frequency

- to give a brighter sound

- to give a "thinner" sounds—this is partly a result of making the sound brighter, and

- to create an effect, whether that be making the filter scream or giving sounds a really squelchy character.

Controlling resonance with an envelope will allow for more subtle nuances to be introduced into a sound.

Filters: Some Sonic Examples

We have talked about filters. Let's build a few patches and listen to the effect of filters in practice.

All of the patches in these first examples are built using Z3TA+. Similar results can be obtained with any of the other featured synthesizers (except for the patches using the formant filter).

This first group of examples are all built around a single sawtooth wave and use some of the filters that are available in Z3TA+—there are more filter options available, both in Z3TA+ and from the other synthesizers featured in this book. Some of the techniques used to build these patches have yet to be discussed, but will be addressed later in the book.

Raw Saw

This isn't difficult .... Raw Saw gives you the sound of a sawtooth wave without any filtering. It is included for comparison with the filtered sounds—it is not the most engaging sound you will ever hear.

Low-Pass Filtering

Now we know what a raw sawtooth wave sounds like, we can start to listen to some filtered waves.

Low-Pass

Low-Pass takes the sawtooth wave and runs it through a 24 dB low-pass filter. As you can hear, the sound is much duller and much quieter than Raw Saw. The volume reduction is not surprising—a significant proportion of the waveform has been removed.

12 dB Low-Pass Sweep

12 dB Low-Pass Sweep puts a sawtooth wave through a 12 dB low-pass filter and then sweeps the cut-off frequency of the filter. A filter "sweep" is another way of saying "adjusts the cut-off frequency from one extreme (for instance fully open) to another extreme (for instance closed)". Filter sweeps can be restricted to a much narrower range and sweeps can also follow rhythmic patterns (see Stuttering Low-Pass Filter below).

At the start of the note, the filter is open and so its effect will not be heard. As the note is held the filter will close down until it reaches a point where it is slightly open and the tone becomes constant. Some resonance has been added to the filter to make the effect of the sweep a bit more extreme (and noticeable), so please ensure the volume is not too loud when you first try this patch.

As you listen to this patch, you will hear the sound is quite bright to start with. As the filter starts to close you will hear the effect of the resonance. Towards the end of the sweep the effect is quite extreme—a few steps further and this patch could be distorting or screaming (turn down your speakers, push up the resonance and listen).

36 dB Low-Pass Sweep

This patch adopts exactly the same parameters as 12 dB Low-Pass Sweep with one difference—the filter is replaced with a 36 dB low-pass filter.

If you compare 36 dB Low-Pass Sweep and 12 dB Low-Pass Sweep, you will hear two main differences. First, with the 36 dB filter the sweep is far smoother and sounds much more controlled: the sound does not come close to screaming or distorting. This smoothness will benefit some patches, but not all—there will always be times when you really want to create a screaming lead sound. Perhaps then you call up a 12 dB filter.

Secondly, you will notice that the sustain sound after the sweep ends is both quieter and more dull. This is a result of using a filter with a sharper slope.

There is one other matter I want to draw to your attention. Listen to the 12 dB Low-Pass Sweep and then 36 dB Low-Pass Sweep—is the effect of the filter in the second patch three times that of the filter in the first? In other words, does a 36 dB/octave filter *sound* three times as effective as the 12 dB/octave filter? Remember, you program with your ears not your eyes. Don't let the specifications fool you about the sonic results. This issue is significant because the 36 dB/octave filter has a higher CPU hit than the 12 dB/octave filter.

Stuttering Low-Pass Filter

Stuttering Low-Pass Filter takes a slightly different approach to filter sweeping and uses a low frequency oscillator to modulate the cut-off frequency of the filter where the frequency of the LFO is linked to the tempo of the track, thereby creating a rhythmic effect. The output level of the LFO is randomized, so the cut-off frequency is randomized too. The effect of these two factors gives the stuttering effect.

High-Pass Filtering

Now that we've listened to a low-pass filter, we can compare that sound to the sound of a high-pass filter.

High-Pass

High-Pass takes the sawtooth wave and runs it through a 24 dB high-pass filter. As you can hear, the sound becomes much thinner and much quieter. While this sound is much thinner than the unfiltered wave, tonally it sounds much closer to the sound of the unfiltered wave (listen to Raw Saw for comparison) than when the wave is run through the low-pass filter. This is not surprising: much of the information about a sound's character is contained in its higher frequencies.

Again, the volume of the sound is considerably reduced when compared with the unfiltered sound because a large chunk of the sound spectrum has been removed.

High-Pass Sweep

With High-Pass Sweep, the sawtooth wave runs through a 24 dB high-pass filter. At the start of the note, the cut-off frequency is low, so as this is a high-pass filter, the effect of the filter will not be heard. As the note is held, the filter sweeps from a low cut-off frequency to a higher cut-off frequency. The sound becomes thinner and quieter.

At the end of the sweep, a constant, quite fizzy tone is heard. If the sweep continued, then all of the frequencies would be filtered and no sound would be heard.

Band-Pass Filtering

Band-Pass filtering combines the characteristics of a low-pass filter and a high-pass filter.

Combo Low + High-Pass

Combo Low + High-Pass is constructed by running two filters in series (ie the output from filter one is fed directly into filter two). The first is a low-pass filter and the second is a high-pass filter. If you compare this patch with Band-Pass, which uses a band-pass filter to create its sound, you will hear that the results are very similar.

Band-Pass

A band-pass filter cuts high and low frequency elements simultaneously and so, as would be expected, with the Band-Pass patch the sound becomes:

- duller

- thinner, and

- even more quiet than when a high-pass or low-pass filter is used on its own.

Band-Pass Sweep

With Band-Pass Sweep, the band-pass filter's cut-off frequency is swept from its highest setting to its lowest setting. At both extremes, no sound is heard due to the filtering effectively removing all elements of the sound spectrum.

Near the top extreme, the sound is reminiscent of a high-pass filter with a high cut-off frequency. Whereas, towards the lower extreme the sound is reminiscent of a low-pass filter with a low cut-off frequency. In between these two extremes you get the effect of the band-pass filter working at different frequencies.

Notch Filtering

A notch filter can give quite subtle results.

Notch

It is often quite difficult to hear when a notch filter is being used. If you compare this patch, Notch, to Raw Saw you will hear that there is a slight thinning of the sound in the mid range. If you played both patches in the context of a track you would be hard pressed to hear the difference.

Notch Sweep

As may be expected, Notch Sweep is constructed by sweeping the cut-off frequency of a notch filter. With this patch you can hear the effect of a filter sweep but without too much energy being robbed from the sound. Where notch filtering is quite a dull and subtle effect, notch filter sweeping can be a far more useful programming technique.

Formant Filtering

Formant filtering can give quite unique results as the following two patches demonstrate.

Formant

The format filter in Z3TA+ effectively has five positions, corresponding to the vowels a, e, i, o and u. This patch, Formant, has been constructed with the format filter set to the "e" position. It is a

41

testament to the power of the filter that a sawtooth wave can at least bear a passing resemblance to the "e" sound.

Formant Sweep

With Formant Sweep, the filter steps through the five vowels—a, e, i, o and u—in turn. This is a long, *long* way from a talking synthesizer. However, the tonal variations are interesting.

Sonic Examples with Different Filters

Those were some sonic examples using Z3TA+. I know want to give you some examples with Surge. The purpose of these examples is to demonstrate:

- the sonic similarities, and

- the sonic differences.

The purpose of this exercise is to listen to the similarities/differences, and not to make any value judgements about the differences. Which sounds right to you will very much depend on the particular track where you want to use the sound.

01 Raw Saw

In order to understand the difference in the sound of the oscillators in the two synthesizers, load up 01 Raw Saw. Listen to the sound on its own, and also listen to the sound when compared with Raw Saw in Z3TA+.

Low-Pass Filtering

Now we know what a raw sawtooth wave sounds like in Surge and have compared it with the raw waveform in Z3TA+, we can start to listen to some filtered waves.

02 Low-Pass

01 Low-Pass takes the sawtooth wave and runs it through a 24 dB low-pass filter. As you can hear, the sound is much duller and much quieter than 01 Raw Saw and has a slightly different character to Low-Pass in Z3TA+.

03 12 dB Low-Pass Sweep

03 12 dB Low-Pass Sweep puts a sawtooth wave through a 12 dB low-pass filter and then sweeps the cut-off frequency of the filter.

With this patch, the sonic differences between Surge and Z3TA+ become more noticeable. Your view may be different (and your view is far more valid than mine since you make your music, and I don't), but to my mind, Z3TA+ sounds more aggressive while Surge is more controlled.

04 36 dB Low-Pass Sweep

This patch adopts exactly the same parameters as 03 12 dB Low-Pass Sweep with one difference—the 12 dB/octave filter is replaced with a 24 dB filter in filter one which is run into a 12 dB filter in filter two giving a 36 dB/octave low-pass filter. Both filters then have their cut-off controlled by the filter envelope.

Clearly this is a more complicated set-up than in Z3TA+ which has a 36 dB/octave filter, however, this set-up does give us a basis for comparison (both with the 12 dB/octave set up and with the sound in Z3TA+).

High-Pass Filtering

Now that we've listened to the low-pass filters, we can compare that sound to the sound of a high-pass filter and can compare how the different high-pass filters sound.

05 High-Pass

05 High-Pass takes the sawtooth wave and runs it through a 24 dB high-pass filter. As you can hear, the sound becomes much thinner and much quieter. You will notice that there is a slight tonal difference between this and the comparator patch in Z3TA+.

06 High-Pass Sweep

With 06 High-Pass Sweep, the sawtooth wave runs through a 24 dB high-pass filter which is then swept with the filter envelope making the sound becomes thinner and quieter as it sustains.

Although this sound is again different from Z3TA+, to my ears it is much harder to distinguish the two swept sounds from one another than it is to distinguish the sounds where the filter is static.

Same Slope: Different Filter Types

We've listened to the differences between Surge and Z3TA+ and I now want to highlight the differences between the different filter types in Surge. Listen to:

- 07 Low-Pass Type 1

- 08 Low-Pass Type 2

- 09 Low-Pass Type 3

These three patches are exactly the same and are based around a sawtooth wave feeding into a 24 dB/octave low-pass filter. However, there is one difference between them: they have different filter types.

As you can hear, the patch with the Type 1 filter has the most obvious, up-front sound which still retains quite a thick tone. The Type 2 filter gives a slightly more restrained sound and may, perhaps, have a slightly thinner tone. The Type 3 filter has a very different character where the resonance is far more muted. This sound might be useful for creating a bass sound for a ballad.

Using Filters When Creating Patches

I will end this chapter with three simple, but usable, patches built using Vanguard. These patches are included as sonic examples of how a filter can be used. As they employ techniques that have yet to be discussed, I will not explain the construction of these patches in great detail (which may make it harder to replicate the sounds if you haven't purchased the patches).

Bass + Stab

This first patch, Bass + Stab, is a straightforward bass patch. The sound is quite full and rich. Having called up the oscillators and closed down the filter (giving quite a dull sound), I then made two main changes to make this patch more playable:

- First, the filter is (slightly) velocity sensitive—hit the keys harder and the sound gets brighter. The filter is also controlled by the volume envelope and so the note is (slightly) brighter in the initial stages.

- Second, in the higher keyboard regions this patch sounded too bright to me so I have applied some key scaling to reduce the brightness of the patch in the top octaves. The key scaling, in conjunction with the filter, works to almost completely cut the sound at higher octaves. With the key scaling the bass patch becomes quite usable as a stab type sound.

Gentle Stab

Keeping with the stab theme, Gentle Stab, uses key scaling but with this patch the sound gets brighter as the patch is played in the higher octaves. This behavior may be more intuitive as it mimics the behavior of natural instruments.

As with the previous patch, the filter is also controlled by an envelope (to give slightly more bite in the attack phase). The filter is also controlled by velocity giving the player real time control over the patch.

There is also another very subtle element to this sound—the resonance of the filter is controlled by the second envelope. The effect of this resonance is to add a touch of brightness to the attack of the note but without dominating the sustain phase.

Resonant Bass

Resonant Bass takes a slightly less subtle approach to the use of resonance. In this patch the filter is controlled by two sources:

- key tracking—as the patch is played at higher pitches, the filter cut-off is reduced making the upper octaves less bright (which would be consistent with a bass patch), and

- velocity—as this patch is played with greater velocity, the filter opens up to give a brighter sound. This gives the player real-time control over the filter.

The resonance of the filter is controlled by an envelope. When the note is struck this envelope works to increase the resonance to the maximum amount and then cuts the resonance. This gives a "squelchy" type sound—the tone of the sound is determined by the filter's cut-off frequency which is controlled by velocity.

You may also notice that in the higher keyboard regions, this patch takes on almost a vocal quality.

Chapter 4
Sound Sources

We've looked at envelopes and we've looked at filters. Both of these have a fundamental effect on the tone of a sound and in many instances can have a greater effect on the perception of a sound than the actual sound source itself. However, the sound source is still significant.

As a first point, don't get too hung up with how different manufacturers label different wave shapes (whether in the six featured synthesizers or with other synths). When viewed on an oscilloscope most waveforms (especially analog or analog-emulating waveforms) do not look like their mathematically generated equivalents. However, while the vintage waves may not look like the mathematically generated waves, they do usually sound great.

As I say, don't concern yourself with labels, but do concern yourself with how individual waves in different synthesizers sound and how you can use each wave in an appropriate context.

All of the synthesizers featured in this book can produce all of the main wave shapes. However, they may produce the waves in different ways, for instance Wusikstation is a sample based synthesizer and so uses samples whereas Cameleon 5000 is an additive synthesizer and so can create a sound from first principles (see Chapter 9: Additive Synthesis for further detail).

This chapter only looks at the main sound sources. It does not attempt to look at all (or even most) of the waves that are available from the featured synthesizers.

Basic Wave Shapes

Let's first look at some of the wave shapes that are common to virtually every synthesizer that has ever been produced.

Sine Wave

The sine wave (see Figure 4.1) is perhaps the most basic element in a sound. It is the purest form of tone you can have—it consists of the fundamental note and has no overtones. If you run a sine wave through a filter, there are no overtones to filter out—therefore the only effect that a filter would have would be to reduce the volume of the note itself. If you put any sound through a low-pass filter, as you take out the harmonics it will tend to sound like a sine wave.

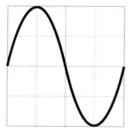

Figure 4.1 The sine wave.

On its own a sine wave can sound quite dull and is not often a first choice for programming. However, as a waveform it is often used to thicken up patches. Where a waveform sounds weak on its own—particularly if it is based on a sampled wave—adding a sine wave can give a depth to a patch and add a roundness/fullness to the sound.

Sine waves are also often added to bass patches to give a sub-sonic, foundation shaking, quality. If you are doing this, please check the patch on full range monitors and take care not to blow your speakers.

A final frequent use for a sine wave is in FM synthesis—it is quite common to build FM patches solely with sine waves (see Chapter 8: Frequency Modulation Synthesis).

As we will discuss later (see Chapter 9: Additive Synthesis), sine waves are the components of all other waves.

Sawtooth Wave

The sawtooth wave (see Figure 4.2) may be the closest that we will come to a general purpose wave.

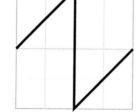

Figure 4.2 The sawtooth wave.
In this case, a rising sawtooth wave.

It gives a bright sound which is often used as the basis for brass and string sounds as well as general "fat" synthesizer sounds (such as stabs and basses). It is rare, but not unknown, to hear raw sawtooth waves—because of the bright quality of the wave it is usually filtered.

While a sawtooth wave is a bright wave, filtering can add warmth and depth to a sound. However, the wave also tends to dominate a broad proportion of the sound spectrum. This may not be a problem, but if your arrangement contains several patches based on sawtooth waves, you may find your mix starts to get muddy.

Square and Pulse Waves

The square waveform has a hollow quality and is often used to create "woody" or "reedy" tones such as those found in woodwind instruments. It is also frequently used in bass sounds, either on its own or to fatten up a sound, often acting as a sub-oscillator (that is a note pitched below the fundamental).

A pulse wave with a value of 50% (in other words when both sides of the wave are balanced) is a square wave. In between the two (as illustrated in Figure 4.3) you get a varying range of tones. A pulse with a 0% width is just noise. Some synthesizers separate the square and pulse, others provide one wave and a facility for pulse-width modulation (which is discussed below).

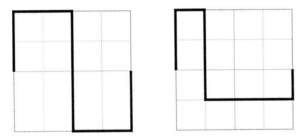

Figure 4.3   A square wave and a pulse wave.

Triangle Wave

A triangle wave (see Figure 4.4) gives a sound that is slightly less reedy, or perhaps less sharp, than a square wave. If you want to stretch a point, you could alternatively think of the triangle wave as being like a sine wave but somewhat sharper in its output (but please do not try to relate a wave's shape on an oscilloscope to its tone).

Figure 4.4   A triangle wave.

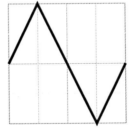

A triangle wave is often used as a low frequency oscillator (LFO) waveform.

Noise

You might think that noise is just noise, however, you would be wrong. It comes in different colors—each color depends on the composition of frequencies in the sound spectrum. I hope you won't think it too much of a cop out if I don't try to use words to describe the sonic differences in great detail. However, in summary, white noise is the brightest, pink noise is slightly less bright and brown noise is duller still.

Complex Wave Shapes and Variations

We've looked at the basic waves. All of the featured synths have many more waves available. Some offer many variations on a basic theme, for instance Z3TA+ gives you 11 sawtooth variants where Surge offers a constantly variable square/sawtooth wave with a sub-oscillator that has a variable pulse-width. This allows you a wide range of sonic options and a detailed level of control. Others simply offer a range of different waves. For all of the featured synthesizers (with the exception of Vanguard) the waveform options are effectively limitless since it is possible to import waveforms.

It would be pointless to try to describe the differences between the waves in words—you really need to spend some time listening to the differences.

You may be wondering why so many waves are offered. There are several reasons:

- different tones and shades—the broader the palette, the more colors the musician can paint with (and the more dilemmas for the sound designer)

- efficiency—some waveforms can be achieved by combining two other waves: by making the wave available without having to use two waves, the developer is giving you the advantage of reduced CPU load and the flexibility to keep other wave slots free

- everyone else does—would you buy a synthesizer which only offered four waveforms (unless it was a very good emulation of a piece of vintage gear)?

However, please do remember that a wide range of waveforms alone does not make a good synthesizer.

Pulse-Width Modulation

Pulse-width modulation (often called PWM) is a technique most associated with square and pulse waves. With a square wave, the positive and the negative phases of the wave are balanced. When the pulse-width is modulated, this balance changes to give a different shaped wave.

The different waveforms are not simply different shapes on an oscilloscope, but contain different spectral components, hence their different tone. These components are discussed in greater detail in Chapter 9: Additive Synthesis.

PWM can either be static, for instance a 50% wave is modulated to give a 40% wave, or it can be a continuous change (for instance when the pulse-width is modulated by an LFO). As a technique, PWM is generally used for one of two reasons. First, it changes the tone of the waveform. Second when two waves which have been modulated in different ways are combined, there can be a fattening effect (but see Doubling Oscillators below).

Wave Shaping

As well as giving a wide range of waveforms, Rhino, Surge, and Z3TA+ also allow you to further shape the waves. The synthesizers take different approaches but the tools all work to distort a raw waveform's shape. The effect of these wave shaping devices can be subtle but can also be more extreme producing FM-like tones (without using FM) and distorted sounds.

With Rhino, if you:

- put the shaper line horizontally, the resulting wave will be a flat line

- put the shaper line diagonally—bottom left to top right—the shaper will have no effect, however

- put the shaper line diagonally—top left to bottom right—the shaper will invert the wave.

Surge offers five waveshaper transform options: soft, hard asymmetrical, sine, and digital. These transformations can be applied with normal or reverse polarity by using the Waveshaper fader. The five transformations get progressively more aggressive; however, they can all create useful musical sounds as well as allowing wilder sonic textures to be generated.

Surge also allows the waves to be twisted when either the Wavetable or Window oscillators are engaged. Figure 4.5 shows the effect of these transformations on a sine wave.

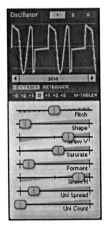

Figure 4.5 Surge's wave transformations applied to a sine wave.

Z3TA+ allows 14 different transformations on its waves. For a more detailed explanation of these transformation, check out my book *Cakewalk Synthesizers: From Presets to Power User* (page 227 and on in that book). Although it doesn't add anything to the sound, one interesting feature of Z3TA+'s shaper is the waveform display (see Figure 4.6) which shows the transformations in real time.

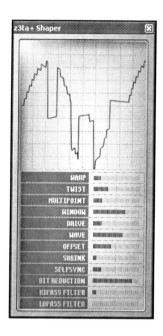

Figure 4.6 The effect of Z3TA+'s shaper on a Vintage Sawtooth wave.

Combining Sounds

You can't have failed to notice that all of the synthesizers featured in this book allow you use more than one waveform at a time and yet so far this book has largely talked about the possibilities offered by using single waveforms.

The reason I have done this is simple—it is easier to explain the operation of single waves. I have already shied away from describing many of the waveforms. If it is hard to find words to describe sounds, it is harder to describe the variation between the many permutations of combinations of oscillators working together. Therefore for the rest of this book there will be a greater reliance on sonic examples than has been the case so far.

There are many reasons for sound combining (including to create new tones, to augment a weak tone, or simply to get a smoother sound) and there are many ways that sounds are combined, for instance two similar sounds could be doubled or two different sounds can be layered to create a wholly new sound.

Whatever the reason, sounds created with multiple oscillators will generally fill more of the sound spectrum. Accordingly, care needs to be taken to ensure that these sounds do not come to dominate a mix (unless that is the intention).

Let's now look at some ways that sounds can be combined in a bit more detail.

Doubling Oscillators

The most simple combination of oscillators is to take one oscillator and clone its settings to a second oscillator. Simply cloning the oscillator may not do much more than increase the volume. However, the addition of a second oscillator does give many more sonic opportunities.

Oscillator Phase

When we talk about the "phase" of an oscillator we generally mean the position in a wave's cycle see Figure 4.7. With one oscillator, the phase matters little, however with two oscillators the effect of phase can be dramatic.

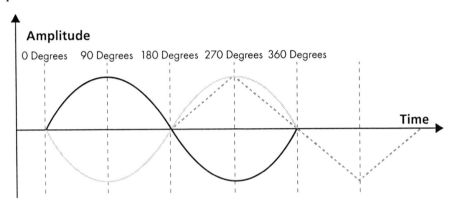

Figure 4.7   The phase of an oscillator. The gray sine wave has reversed polarity (when compared to the black sine wave) and the dotted triangle wave is 180 degrees out of phase.

If you have two waveforms that are totally in phase then you will get reinforcement of the signal. Reinforcement is perceived as an increase in volume. If your waves are out of phase then you will get cancellation. The effect of the cancellation depends on the individual waves: if you have two sine waves that are 180 degrees out of phase, then you will get total cancellation.

Cancellation caused by putting waves out of phase is generally perceived as a change in tone, usually making the sound thinner and sharper. This may be a great result if you are after a plucked sound or a more metallic sound—it may be less impressive if you are after a really fat sound.

If we look at Z3TA+ it gives us several options to play with the phase which may be a useful illustration of the sonic possibilities of phase. In Z3TA+ all of these changes can be applied on a per oscillator basis (so each of the six oscillators can have different combinations of options):

■   The phase of a note can be synchronized with the start of a note (in other words, the key strike). This means that every note from that oscillator will start from the same position in its phase—the starting phase position will be determined by the Phase control for that oscillator.

- The phase of each note can be synchronized but the polarity reversed—mix a wave with positive polarity and a wave with negative polarity, both with the same phase, and there will be total cancellation.

- Alternatively the phase of each note (again on a per oscillator basis) can be allowed to run freely. This means that each new note has a different phase and each note from a different oscillator can have a different phase. This can result in some interesting effects (or annoying effects, depending on your perspective).

- As mentioned earlier, the phase of each note can be shifted—this can give consistency of tone when using two oscillators each of which is key strike phase synchronized.

As you will hear, the effect of detuning an oscillator can lessen the more extreme effects of phase shifts.

One thing to note if you are playing with the phase in Z3TA+ is that it does not make phase changes in real time. Instead, the new phase is introduced with each newly triggered note.

Oscillator Detuning

Another common technique used with doubled oscillators is to detune one of the oscillators (or to detune each of them but in different directions). Subtle detuning can give a natural chorusing effect which is perfect for creating fullness/roundness/smoothness or just adding fatness to the sound.

There is a balance to be struck when using this technique—if the detuning goes too far then the resulting sound can become flabby and/or out of tune. This may be a good or bad thing, depending on the effect you're after.

Rhino, Surge, Wusikstation, and Z3TA+ all allow you to fine tune individual oscillators. Cameleon 5000 does not allow for detuning of individual oscillators, however, each partial can be individually detuned (detune them all and you *might* get a similar effect). Vanguard doesn't allow this fine detuning as such (but see below for how it addresses this issue).

Multi-Oscillators

So far we have looked at doubling oscillators. However, there is no reason why you shouldn't triple or quadruple oscillators and separately detune each oscillator. Four of the synthesizers offer six oscillator slots which would allow you to build a patch with six slightly detuned oscillators.

Or you could use one of the multi-oscillators that are available in Surge, Vanguard, Wusikstation, and Z3TA+ and in one step save yourself a lot of time and perhaps get a better sound.

While different in their implementation, all four have the same essential characteristics. Instead of there being one wave, with the multi-oscillators up to 16 waves (depending on the synthesizer) will be called up (in one slot). These waves will be spread across the stereo spectrum and slightly detuned to give a very big sound from one oscillator.

The naming of these option is different with each of the four synthesizers, as is the range of control offered.

- Surge allows the number of waves to be selected (up to sixteen) and the amount of detuning to be controlled, both on a per-oscillator basis. In addition, Surge has an Oscillator Drift fader which introduces an element of random tuning (much like a genuine analog oscillator) in order to give a less static sound.

- Vanguard allows control over the number of voices (up to ten) and the amount of detuning. In addition, an Age control introduces some analog-emulating drift.

- Wusikstation has a master unison option (which, like Vanguard, applies to all oscillators). With this option the amount of detuning can be selected from ten options.

- Z3TA+ has a combined detune/voice count control. It also offers the option of having the phase of all of the detuned voices synchronized or not. Like Surge and Vanguard, there is a Random slider to introduce some old-fashioned pitch drift.

Layering Oscillators

There is a fine difference between layered and doubled oscillators—I am only using different terms so that I can be clear about the different concepts.

With layered oscillators, you again take two (or more) oscillators acting together. However, the oscillators are different, so for instance, you may combine a saw and a square wave. Alternatively you may combine two square waves, one being an octave higher than the other.

The combinations may or may not be detuned—that is all a matter of taste as we shall hear later in this chapter. However, it is quite common to change the octave of a layered oscillator so that it can add some high end bite or low end punch.

For a simple practical example of layered oscillators, take a sawtooth wave and a square wave. On their own, both waves have a certain sound: both are quite bright and in certain circumstances this may mean that some richness and some depth of tone is lacking. You could double either of these oscillators to get a richer tone. Alternatively, you could layer these two oscillators.

Layering gives a new tone that is neither sawtooth nor square—the layered sound still has the brightness of the two components, but perhaps more weight. If you then want to thicken things up considerably, drop the square by an octave and engage the multi-mode oscillator for the sawtooth and put the result through a filter. This is a very quick and dirty way to get a fat sound.

Take a look at the figures on the next page. Figure 4.8 shows two sine waves being added: one being pitch an octave above the second. Figure 4.9 shows a sawtooth wave and a square wave (both of the same pitch) being added together.

Layering Sounds

If there was a fine difference between layered oscillators and doubled oscillators, the difference between layered oscillators and layered sounds is even more tenuous.

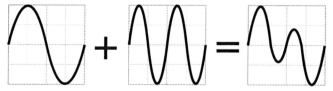

Figure 4.8   When a sine wave, and a sine wave pitch an octave higher than the first are added together, the result is a wholly new wave.

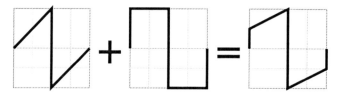

Figure 4.9   When a sawtooth wave and a square wave are added together, the result looks like a combination of the two.

With layered oscillators you take two (or more) oscillators and route them together through the same modifiers and controls (filters, envelopes etc)—in effect, the two oscillators are working together as one, creating a whole new sound.

With layered sounds, each oscillator has its own filter and its own envelopes etc. The advantage of a layered sound (when each layer has a similar sound) over a layered (or doubled) oscillator is that you can get a thicker, fuller sound which is more controllable. However, one of the advantages of layered oscillators is the speed of working for nearly the same sound.

As with doubling and layered oscillators, layered sounds can be detuned.

Whether you can layer the sounds or layer the oscillators depends on the particular synthesizer's architecture. For instance with Vanguard you can only layer (or double) the oscillators but with Wusikstation you can only layer the sounds. Rhino, Surge, and Z3TA+ allow you to layer both (or either) sounds and oscillators—indeed, Surge is effectively two synthesizers bolted together which share some FX units. Neither option is right or wrong (or better or worse)—its just how the synthesizers have been designed.

Layering sounds is more of a programming technique rather than being a separate sound source and so there are no sonic examples with this chapter. However, the technique will be demonstrated in Chapter 12: Building Patches. Also, as you will see in Chapter 7: Frequency Modulation Synthesis, layering is one of the key building blocks for FM sounds.

You may come across synthesizers that offer "vector" synthesis (for instance Wusikstation). Vector synthesis is another way of layering and fading between sounds. With the original hardware vector synthesizers there was a joystick that gave the sound designer control over the mix of a number of sounds (usually four). The days of hardware are largely gone, but some software synthesizers (for instance Wusikstation and Z3TA+) still allow this kind of control to mix sounds together.

Syncing Oscillators

We've already discussed the phase of oscillators. There is a further step you can take (in Z3TA+) and hard synchronize two oscillators together. When hard sync is engaged, the slave oscillator restarts its phase each time the master oscillator starts its phase.

If the two oscillators are pitched at the same level, the effect is comparatively mild (perhaps giving some cancellation). If the two oscillators are pitched differently then one oscillator will complete its phase before the other—this means that the slave will be part way through its phase when it is retriggered. This can result in a very hard sound that is often used for creating cutting lead sounds. You can achieve a similar effect with the Classic oscillator in Surge.

Combining Sounds: Creating New Tones

That's enough theory, let's look at some practical patches. These patches have all been built with Z3TA+ since this synth has the most easily accessible flexibility for the points I am trying to demonstrate, and if we stick with one machine we can make a consistent comparison.

We're going to look at—or rather listen to—the effect of combining two sounds. This is much like the musical equivalent of paint color charts when you want to redecorate your home. On their own each color looks fine—it is only when you put the color next to a similar color that you can see the similarities and the differences. And as with color charts, when you have a range of tones, there is a dilemma about choosing the right one.

In listening to these combinations, we're trying to achieve several things:

- first to hear the added weight of combining two oscillators

- second to hear a completely new tone, and

- third to hear whether the component parts can be separately identified.

We're also going to listen to how the combinations react to filtering.

The comments in this section relate solely to Z3TA+—the sounds which are created are based on the interaction of Z3TA+'s oscillators and filters. While many comments may be equally applicable to many other synthesizers, these views should not be regarded as "rules" and the interaction of the same elements in different synthesizers may give differing results.

Sawtooth Waves—Same Octave

The next two patches are included as a reference point for later comparisons.

Single Saw and Single Saw Filtered

Single Saw gives the sound of a single sawtooth waveform. Single Saw Filtered gives a single sawtooth wave filtered through a 24 dB/octave filter which is velocity sensitive (as are all of the filtered patches in this section). Velocity sensitivity means that the filter's cut-off frequency varies according to the MIDI velocity (that is the loudness) of the incoming note.

On its own, the sawtooth wave sounds rich and bright. Through a filter, there is a progressive cut in the brightness of the wave.

Two Saw Running Free

With Two Saw Running Free you can hear two sawtooth waves running free—in other words, the phase of each of the two waves is arbitrary. This means the waves could both be in sync or could be totally out of sync with reference to each other.

If you strike the same note repeatedly, you will hear different tones caused by the differing phase cancellation of the two oscillators.

Two Saw Phase Sync

In Two Saw Phase Sync, the phase of each of the oscillators is synchronized to the key strike. Because of the synchronization, this patch has a consistency that is not heard in Two Saw Running Free: each time you strike the key you will get exactly the same note. As both oscillators are running exactly in sync, the effect is to increase the volume output without changing the tone.

Two Saw 90 Phase

Two Saw 90 Phase has the same set up as Two Saw Phase Sync, however, the phase of oscillator two is 90 degrees behind oscillator one. This means that there will be some reinforcement and some cancellation which together result in a different tone. You can hear that this tone is much thinner and has more of a plucked quality—perhaps the sort of sound that could be used as the basis for a harpsichord type patch.

As the phases of the two waves are synchronized there will be a consistent tone to this note. It would have been possible to have two free-running oscillators and to put one 90 degrees out of phase by reference to the other. However, that would have been a fairly pointless exercise as the waves would have been free-running in the first place (a random phase that is 90 degrees different is still a random phase).

Two Saw 180 Phase

Two Saw 180 Phase takes the previous patch and pushes the Phase slider in the second oscillator to 180 degrees. The effect—especially in the last few degrees—is radical. A thin plucked tone loses even more of its bass element to end up sounding as if the patch has been raised by an octave. Indeed, if you drop this patch by an octave and compare it with Two Saw Phase Sync you will hear similarities between the sounds. Figure 4.10 illustrates the resulting wave.

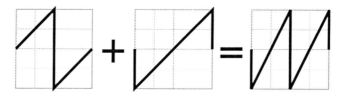

Figure 4.10   By combining two sawtooth waves—one of which is 180 degrees out of phase with the other—the result is a sawtooth wave, but raised by one octave.

Two Saw Inverted 1 Phase

If we take two sawtooth waves and invert them without any phase differential there would be total cancellation. I guess that you know what silence sounds like and so wouldn't appreciate a demonstration. Instead with Two Saw Inverted 1 Phase, the second wave is inverted by reference to the first and its phase is moved by one degree to ensure there isn't total cancellation.

You can hear that there is still considerable cancellation, and the sound that is audible has little tone and a lot of noise. If you move the Phase fader the tone will shift, gradually becoming richer. If you move the fader to 180 degrees you will hear another shift in tone with the sound of a square wave becoming audible. This is not surprising—if you add two sawtooth waves together with one having an inverted polarity and being 180 degrees out of phase, the result will be a square wave. Figure 4.11 shows this concept in practice.

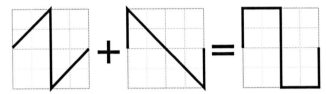

Figure 4.11   By combining two sawtooth waves—one with its polarity reversed and 180 degrees out of phase—the result is a square wave.

Two Saw Detune Running

Two Saw Detune Running takes two sawtooth waves, both with the same polarity, but neither synchronized, and slightly detunes the first oscillator.

As you can hear, especially if you compare the patch with Two Saw Running Free, the detuning results in a much warmer tone which is constantly shifting. Where Two Saw Running Free may not have provided many usable tones, this patch could find many uses.

While there have been no changes to the envelope settings, this patch almost appears to fade in (especially when compared with the next patch).

Two Saw Detune Phase Sync

Two Saw Detune Phase Sync takes the previous patch and synchronizes the phase of the oscillators when a key is struck. This gives three main differences when compared with Two Saw Detune Running:

- the tone of this patch is different having a slightly brighter quality (the comparison obviously depends on how closely synchronized the oscillators are at any stage in Two Saw Detune Running)

- the texture of this patch is different—with the phase start synchronized, the resulting sound has a far more aggressive quality, and

- there is a real "spit" to this patch in the attack phase by comparison with Two Saw Detune Running.

The differences between Two Saw Detune Running and Two Saw Detune Phase Sync are significant when building patches. The free-running patch may give a sound that is far more suited to a soft pad where the phase-synchronized sound may have more of a use in a stab type sound.

Two Saw Big Detune Phase Sync

Developing the previous patch further, Two Saw Big Detune Phase Sync takes Two Saw Detune Phase Sync and shifts the tuning of oscillator one, putting it 50 cents out of tune with oscillator two.

The result, to my ears at least, is a not very usable tone. However, even if we can agree that this sound is not particularly usable, I think we could probably find a range of opinions disagreeing about where the useful tone ends and the "out of tune" sound begins.

Two Saw Detune 90 Phase and Two Saw Detune 180 Phase Sync

In the same way that Two Saw 90 Phase and Two Saw 180 Phase have a thinner sound than Two Saw Phase Sync, these two patches have a thinner sound than Two Saw Detune Phase Sync, but they obviously have a fuller, richer sound than their comparators with equally tuned oscillators. There is not quite such a marked tonal difference when the waves are 180 degrees out of phase, however, if you adjust the tuning, as the waves get closer in pitch the thinning effect becomes far more pronounced.

You can also hear that as the phase difference gets closer to 180 degrees, some of the edge gets taken off the attack.

Two Saw Detune Inverted

If you simply invert the polarity of a wave and then play it with a normal polarity version of the same wave, you will get total cancellation. This is not the case when two opposite polarity, slightly detuned waves are played.

However, you do get quite a thin sound and irrespective of the envelope settings, the sound will fade-in slowly—if you adjust the fine tuning, the closer the two oscillators are pitched together, the longer the fade in time.

Saw Multi Free Filter

Taking a slightly different approach, with Saw Multi Free Filter we have:

- a single oscillator but with multi-mode called into action, in other words you are not listing to one oscillator but many voices, each slightly detuned

- the many voices created by the multi-mode oscillator do not have their phases in sync, and

- the output then passes through a 24 dB/octave low-pass filter which is controllable with velocity.

If you compare the sound of this patch with Two Saw Running Free you will hear a very different tone—this patch is much richer and doesn't have the sharpness associated with phase cancellation that is present in the two saw patch (although there will be phase cancellation in this patch).

If you experiment with the velocity you will also hear that the filter will smoothly control the tone of the oscillator.

2 × Saw Multi Detuned

Building on this, 2 × Saw Multi Detuned uses two multi-mode sawtooth waves—one slightly detuned and both with the waves synchronized to the key strike. No filtering is used in this patch.

You will notice several different characteristics to this sound:

- most immediately, there is a real spit to this patch—that is a function of having all of the waves synchronized to the key strike, and

- second the tone is much harder (even though there are two oscillators and therefore double the number of waves. There is also (slightly) less movement in the tone.

To my mind there may not always be much advantage to using two multi-mode oscillators instead of one. I'm not suggesting you should not use two multi-mode oscillators, just that the tonal differences may be quite subtle.

So What?

We've listened to some sawtooth waves in various combinations. What have we learned from this? With sawtooth waves, there seem to be a few important principles:

- Two, slightly detuned oscillators give a bigger sound than one.

- Keeping two oscillators at the same pitch gives a sharper sound, especially when the two waves' phases are synchronized.

- Synchronizing the phase starts gives a much harder quality to the sound.

- Synchronizing the start of two waves and then starting their phases at different stages in the cycle generally gives a thinner sound.

- Too much difference in the tuning and the sound starts to lose focus and "sounds" detuned.

- 1 sawtooth wave + 1 sawtooth wave = 1 square wave (provided you understand phase and polarity).

Square Waves—Same Octave

Let's look at some square waves and see if the principles are similar to those we have found for sawtooth waves.

Single Squ and Single Squ Filtered

Like their single saw equivalents, these patches are here for comparison purposes.

Single Squ gives the sound of a single square waveform. Single Squ Filtered gives a single square wave filtered through a 24 dB/octave filter which is velocity sensitive.

On its own, the square wave sounds bright with a certain sharpness. However, when listening to this sound through a filter there is not much of a progressive change to the square wave's sound—it is either bright or dull: there is little in between.

Two Squ Running Free

Because square waves are either at their positive maximum or at their negative maximum, phase cancellation has quite a stark effect. With Two Squ Running Free, when the two waves are not in phase, the sound can take on an FM-like, almost metallic type quality.

Two Squ Phase Sync

In Two Squ Phase Sync, the phase of each of the oscillators is synchronized to the key strike. Because of the synchronization, this patch has a consistency that is not present in Two Squ Running Free: each time you strike the key you will get exactly the same note. As both oscillators are running exactly in sync the effect is to increase the volume output without changing the tone.

Two Squ 90 Phase

With Two Squ 90 Phase, the phase of the waves is synchronized to the key strike, but the phase of the second wave is 90 degrees behind the first wave. While there is a different tone from Two Squ Phase Sync—a thinning of the sound—there isn't the same difference in tone as occurred in the sawtooth version of this patch.

Two Squ 179 Phase

If a square wave is added to another square wave which is 180 degrees out of phase, it has the same effect as mixing two square waves which have opposing polarities.

Again adopting the logic that everyone knows what silence sounds like, this combination mixes two square waves, the second of which is 179 degrees out of phase. The result is a very thin and noisy sound: there is also a lot of cancellation going on so the sound is very quiet.

If you check out some of the different tones when different phases are selected, you will notice that the tone does not vary as much as it does if you do a similar exercise with sawtooth waves. While the sound does get thinner, its essential tone remains.

Two Squ Detune Running

Two Squ Detune Running takes two square waves with free-running phase and slightly detunes one. The sound produced by this patch is a square wave with chorusing.

Two Squ Detune Phase Sync

Two Squ Detune Phase Sync builds on the last patch and synchronizes the phase of both oscillators to the striking of the key. The note has a natural chorusing effect, but the sound is much harder and the attack of the note is far more pronounced.

Two Squ Big Detune Phase Sync

When the pitch difference between the two phase synchronized oscillators is increased to 50 cents, the chorusing effect becomes unpleasant and is perceived as dissonance. If you add a slow LFO modulating the pitch of both oscillators simultaneously you might be able to use this tone to create a siren effect—if you have purchased the patches to accompany this book, push the mod wheel up a bit you will hear that this effect has been programmed.

Two Squ Detune 90 Phase

Unsurprisingly, Two Squ Detune 90 Phase sounds very much like Two Squ Phase Sync but with some added natural chorusing effect. And like its non-detuned counterpart, the phase difference has little effect on the tone, in fact, with the detuning it is even harder to tell the tonal difference between this patch and Two Squ Detune Phase Sync. If you do a direct A/B comparison, you will hear the difference. I suspect that within the context of a patch or mixed in a track the difference would not be noticed.

Two Squ Detune Inverted

If you take two waves both having their phase key synchronized but with their polarity inverted, you would expect there to be total cancellation. However, in this patch one wave is slightly de-tuned and so there isn't complete cancellation. Instead the attack portion of the wave sounds as if the attack has been slowed so that the patch fades in.

Square Multi Free Filter

As we did with the sawtooth waves, Square Multi Free Filter is based on:

- a single oscillator with multi–mode engaged

- all of the waves created by the multi–mode oscillator having their phase in sync, and

- the output then passing through a 24 dB/octave low-pass filter which is controllable with ve-locity.

To my mind a square wave on its own gives a very sharp sound. However, when multi-mode is engaged, the quality of the sound changes completely, almost giving a vocal-type quality to the sound: put the wave through the formant filter to hear what I mean.

What I find particularly interesting about this patch is how the multi-mode wave reacts to the fil-ter. Generally I find that with a square wave the filter is hard to use—the filter either has too little effect or it has too much effect (and even then you seem to get the worst of both worlds—a dull characterless tone that is still too angular).

I much prefer how the multi-mode square wave reacts to a filter: there is a progressive (and musi-cally useful) change in the tone as the filter closes. In other words, the combination works just as you would hope.

2 × Squ Multi Detuned

2 × Squ Multi Detuned uses two multi–mode square waves—one slightly detuned and both with the waves synchronized to the key strike. No filtering is used in this patch.

As might be expected the tone of this patch is much harder and there is also less movement in the sound. However, if you do call up a filter, it works as well as for a single multi-mode square wave oscillator.

Sawtooth Waves—Octave Separation

So far we have looked at the basic tones and only considered the effect of the filter in a few instances. We are now going to look at more complex combinations and listen to how the filter works with these combinations.

To try to keep the permutations within the range of human tolerance, all of the following group of patches have the oscillators' phases synchronized to the key strike.

These patches are all based on two waves tuned to different octave intervals. The octave intervals were chosen as they are the most easily usable and to try to reduce the number of permutations. You should feel free to experiment with all and any interval differences.

Two Saw +1 Octave

This first patch takes two sawtooth waves—the second oscillator is tuned an octave higher than the first.

The effect of the second oscillator on the tone is radical when compared with Two Saw Phase Sync. You no longer hear the sound of a sawtooth wave, but instead you hear something brighter and much sharper—almost with a plucked tone, perhaps beginning to resemble a harpsichord wave or similar.

As you can hear, adding two bright but not exceptional waves together gives us a completely different tone which has no resemblance to the component parts.

Two Saw Detune +1 Octave

Taking the previous patch a step forward, Two Saw Detune +1 Octave slightly detunes the base oscillator (the effect is the same whichever oscillator is detuned). The detuning gives a chorusing effect as occurred when the oscillators where both at the base pitch. However, this chorusing effect is much more subtle—more in the way of a warming/thickening of the tone.

Two Saw +1 Filter Octave

Two Saw +1 Filter Octave, takes Two Saw +1 Octave (which has no detuning) and runs both waves together through a 24 dB/octave low-pass filter. The filter cut-off frequency is controlled by velocity.

While the waveform of the combined oscillators without filtering sounds nothing like a sawtooth wave, the combined waves react to filtering in the same favorable way that individual sawtooth waves do, in other words there is a constant shaping of the tone and a range of tone colors can be achieved.

You can also hear that with more extreme filtering the resulting sound is close to the sound of a similarly filtered single sawtooth wave.

Two Saw +2 Octaves

When two sawtooth waves are tuned two octaves apart, the sound takes on a brighter and thinner tone still. There is still one distinct unified sound—you do not perceive this to be two waves playing together at the same time.

The difference between this patch and Two Saw +1 Octave is noticeable, however it is nowhere near as radical as the difference between Two Saw +1 Octave and Two Saw Phase Sync.

Two Saw Detune +2 Octaves

With a slight bit of detuning, the sound of the combined waveform in Two Saw Detune +2 Octaves is quite clear but the detuning adds a bit of brightness and thickness to the tone. It would be hard to describe this sound as natural chorusing.

If the amount of detuning is increased, the effect is to make the sound take on more of a "drone" and for the two sound sources to become individually identifiable.

Two Saw +2 Filter Octaves

As may be expected, when the two sawtooth waves separated by two octaves are put through a filter, the sound again tends to become reminiscent of a filtered single sawtooth wave, but as it is slightly brighter, it also sounds slightly thinner. With less extreme filtering, the sound reacts well to filtering providing a broad range of changing tones.

Two Saw +3 Octaves

With Two Saw +3 Octaves I have raised the second oscillator by another octave so that there is a three octave gap between the oscillators. Two separate tones can now be clearly heard. The sound has a quite distinct bright organ-like quality with a lower tone and a higher tone being heard simultaneously.

Two Saw Detune +3 Octaves

As you can hear, when two sawtooth waves which are three octaves apart are combined and one wave slightly detuned, the detuning has little effect on the sound. For this reasons we won't listen to the difference with larger octave intervals as the tonal effects become meaningless.

However, if you detune one oscillator significantly (say by 50 cents) then the effect becomes quite unpleasant.

Two Saw +3 Filter Octaves

If you liked the combination of the raw waveforms in Two Saw +3 Octaves you are more likely to appreciate the filtered version—if you didn't, you may not like the filtered version. As with all of the sawtooth based combinations, the filter is quite effective, and as the filter closes the sound becomes more like a filtered single sawtooth. While the filter is quite effective, it removes the effect of two sounds by filtering out the higher sound.

Since we are using a low-pass filter, as the filter closes it has more effect on the higher pitched oscillator. This means that it cuts the higher pitched oscillator disproportionately leaving the regularly pitched oscillator to dominate. If you are modulating a filter with an envelope this could mean that the higher oscillator can be heard during the attack phase with only the regularly pitched oscillator sounding during the sustain phase. This combination could be used to give a regular sawtooth sound more bite. This effect is demonstrated in Harpsi Two Saws +2 later in this chapter.

While a similar principle is true for smaller intervals, the effect is less noticeable.

Two Saw +4 Octaves

With the interval set to four octaves, the resulting sound of Two Saw +4 Octaves is quite sharp and buzzy. I'm not sure that you would want to use this waveform combination without filtering.

Two Saw +4 Filter Octaves

Many of the principles that applied in Two Saw +3 Filter Octaves are relevant here. Again, the raw tone combination may not always be very useful, however as an example, with an envelope modulating the filter you may be able to create an organ-type click on at the start of a note.

Compare A and Compare B

Before we move on I want to highlight some of Z3TA+'s more useful waveforms.

Listen to Compare A and Compare B and compare the sounds. Do you notice any difference?

To my mind they are very close although I think Compare B has a more cohesive tone than Compare A (in other words you hear only one sound, not two) and also Compare B has a slightly thinner sound than compare A.

However, there is a far more practical difference between these two patches:

- Compare A is made up of two Vintage Saw 1 waves, one pitched three octaves above the second and at a lower volume (35% against 60% for the base oscillator, so there is a different balance from that found in Two Saw +3 Octaves).

- Compare B only uses one wave—Octaved Saw 2.

There are some reasons why you might want to use Compare A:

- you prefer the tone (unlikely), or

- you want the flexibility to control each wave separately whether in terms of pitch or volume (for instance so that you can mix them in different proportions or change the levels individually over time).

However, there are three reasons why you might want to use Compare B:

- a single oscillator uses less CPU

- using one oscillator leaves more oscillator slots free for further sonic possibilities, or

- you prefer the tone of the combined wave.

You may find that Z3TA+ doesn't offer the wave combination you want. You can still take advantage of this sort of combined waveform by calling up the waves you want to combine, sampling the output of the combined wave (you can do this by rendering a wave in your host) and then importing the sample back into Z3TA+.

Combining Sawtooth Waves—A Few Thoughts

We have now been through a large chunk of sawtooth wave combinations. There are a few thoughts that we can draw from what we have heard:

- combinations of waves pitched at different octaves give radically different tones

- sawtooth waves will still react to a filter when they are combined with other waves, and

- you don't necessarily need FM or additive synthesis to get brighter tones. This is not to say that you can dispense with FM and additive—the combinations shown here are used in a subtractive synthesis context and do not necessarily offer the full range of sonic possibilities which are available with additive synthesis and other sound creation methods not based around tone shaping with a filter.

Square Waves—Octave Separation

We've looked at the sounds produced by combinations of sawtooth waves at different octaves. We're now going to undertake a similar exercise for the square waves and listen to the new tone variations that these combinations offer.

Two Squ +1 Octave

Where two square waves at the same pitch will give quite a hollow, wooden tone, pitching one square wave an octave higher than the other gives the combination a much thinner sound with some of the characteristics of a sawtooth wave but somewhat brighter. If you drop the level of oscillator two to about 25% you will hear the sawtooth sound more clearly.

Tonally the combination of these two square waves gives a new sound—the individual elements cannot be separately identified.

Two Sq Detune +1 Octave

If you listen to and compare Two Squ +1 Octave and Two Sq Detune +1 Octave you will hear very little difference—there is perhaps a slight change in the tone, but certainly none of the chorusing effect that we have come to associate with detuning. The effect is nowhere near as dramatic as it is for the corresponding patch constructed with sawtooth waves nor as dramatic as it is when both the square waves are evenly pitched.

At higher octave separations, the detuning effect becomes even less significant so I haven't included any further examples of combinations of detuned square waves. However, do feel free to experiment on your own.

Two Squ +1 Filter Octave

As I mentioned earlier, I am not a great fan of filtering square waves. The effect of the filter when one square wave is pitched an octave above another gives a better filtering result than when a single square wave is filtered. However, I'm still not knocked out by the results.

With Two Squ +1 Filter Octave the filter does work to progressively change the tone, however, in the same way that when square waves are filtered the essential tone does not change that much, with certain filter settings you can still get the effect of a waveform that is dull but with too many overtones.

Two Squ +2 Octave

With the second square wave being tuned two octaves above the base oscillator, you start to hear two tones although the effect is not noticeable to a distracting point. As you would expect, the sound is brighter and even thinner, getting still further away from the tone of a square wave.

Two Squ +2 Filter Octave

Two Squ +2 Filter Octave reacts well to a filter (subject to my usual comments about the difficulties when filtering square waves). You do need to exercise some caution when filtering this combination as it has a ready tendency to sound like an organ.

However, this patch (along with Two Squ +3 Filter Octave) does illustrate an important point for additive waves with dominant harmonic content where you will find similar difficulties if you want to filter the wave.

Two Squ +3 Octave and Two Squ +4 Octave

When one square wave is raised by three or four octaves, two sounds can be clearly heard. The sound is brighter than if you listen only to the base oscillator, however, as the two sounds remain distinct, the tone does not get thinner through adding the higher pitched oscillator.

The key difference between Two Squ +3 Octave and Two Squ +4 Octave is the quality of the higher pitched element—as might be expected, with the higher pitched second wave, the tone of the patch is much sharper.

Two Squ +3 Filter Octave and Two Squ +4 Filter Octave

As we found with earlier examples using two oscillators with a large interval between their tunings, when passing both oscillators through a low-pass filter, as the filter closes it has more effect on the higher pitched oscillator. This means that it cuts the higher pitched oscillator disproportionately leaving the regularly pitched oscillator to dominate.

If you are modulating a filter with an envelope you could get the higher oscillator to sound during the attack phase with only the regularly pitched oscillator being audible during the sustain phase. This combination could be used to simulate the key click of an organ—the quality of the click could be adjusted by changing the tuning of the higher pitched oscillator.

Combining Sawtooth and Square Waves

We've listened to combinations of sawtooth waves and combinations of square waves separately and so now we're going to listen to combinations of the two waves working together. For this group of examples we are going to listen to the effect when the sawtooth and the square wave work together and in particular, the effect when one is pitched higher than the other.

Saw Squ

The first patch we will listen to takes a sawtooth wave and a square wave both at the same pitch. The effect is to create a wholly new sound which is:

- brighter than a sawtooth alone
- fuller than a square wave alone, and
- more aggressive than either.

So is this the ideal wave combination? Perhaps. Perhaps not. It all depends on how you want to use the combination and whether you agree with my assessment of the sound. However, if you are trying to get a big, dominating sound this combination works well and is a good starting point.

Saw Squ Detune

When the sawtooth wave and the square wave are used in combination but with a slight detune, there is some natural chorusing, but not a significant amount.

Saw Squ Filter

The combination of sawtooth wave and square wave reacts well to being filtered, giving a broad range of tones. However, the tone of the square wave does tend to become clearer as the filter is closed.

Saw +1 Octave Squ

When the sawtooth wave is raised to be an octave higher than the square wave the result is a thicker tone than when the two oscillators are pitched together.

My preference—and you may disagree—is not to mix these two waves in equal measures. When you allow one wave to dominate:

- if the square wave is the quieter, the tone of the sawtooth can predominate with the square wave being used to fatten out the sound, but
- if the sawtooth is the quieter, the square wave can predominate but some of its harshness is counteracted by the sawtooth.

Saw Squ+1 Octave

By contrast, when the square wave is raised an octave above the sawtooth wave, the result is a thinner tone than the tone when the sawtooth is higher. When the volumes of these levels is changed so that one is quieter than the other, the louder wave tends to dominate and the quieter wave adds little to the tone.

Saw +1 Squ Detune and Saw Squ +1 Detune

There is a very slight effect when one of the oscillators is detuned against the other. When the sawtooth wave is an octave higher, the effect is more in the nature of a slight tone change. When the square wave is higher, the effect is more of a chorusing effect.

With higher octave intervals the detuning has progressively less effect and so I won't look at any further examples of detuning in this section.

Saw +1 Squ Filter and Saw Squ +1 Filter

Both of these patches react well to the filter and allow a progressive tone change without either wave coming to dominate. You will notice that as the filter is closed it becomes harder to differentiate between these two sounds.

Saw Multi +1 Squ and Saw Multi +1 Squ Filter

For the final patch of the one octave separation group, I have taken a multi-mode sawtooth wave and combined it with a single square wave (set at a lower volume). For the second patch, I have run this combination through a 24 dB/octave filter.

For both patches the square wave works to really thicken up the multi-mode sawtooth giving some guttural punch to the patch.

However, the combination of the two waves is not quite as seamless as was the case when neither wave engaged multi-mode (listen to Saw +1 Octave Squ and Saw Squ +1 Octave for a comparison). In Saw Multi +1 Squ and Saw Multi +1 Squ Filter the square wave can be heard as a separate component to the sound—it is almost a low level buzz. This effect is less pronounced if you use the patch in a track, however if you just hold a single note, you can hear the effect quite clearly. You can also try this patch with a multi-mode square wave. To my ears, engaging the multi-mode square loses the effect of the square wave and does not add much to the patch.

This combination does filter well and as the filter closes, the sound takes a more cohesive tone with the low level buzz of the square wave still being present but being less obvious as the contrasting sawtooth wave is reduced.

Saw +2 Octave Squ and Saw Squ+2 Octave

When one oscillator is pitched two octaves above the other, the two waves can start to be heard as distinct elements. The tone gets brighter but this is more a reflection of the dominance of the higher note. As the two elements are being heard separately, there is little change to the tone when their respective levels are shifted.

Saw +2 Squ Filter and Saw Squ +2 Filter

Both of these patches react well to the filter. When the filter is closed there is a cohesive tone (rather than there being two separate sound sources). When the square wave is pitched higher, the effect of the filter is less significant for the tone—it readily cuts the sawtooth element but only has a dulling effect on the square wave.

Saw +3 Octave Squ, Saw Squ+3 Octave, Saw +3 Squ Filter and Saw Squ +3 Filter

When a sawtooth wave and a square wave are pitched three octaves apart, the sound starts to become less useful as a singular tone and is much more a combination of two elements. That being said, the combination does give some good organ type tones and the sound does react quite well to filtering.

So Why Have We Listened to All These Combinations of Waves?

We have taken two waves and gone through around 70 permutations of how those waves can be combined. Why?

There are many reasons—the first was to introduce you to an element of the sound palette that is available—you need to take some time to listen to them to become aware of the possibilities. Given the components, some of the tones and colors are neither obvious nor intuitive.

Once you are aware that there is a range of tones beyond those offered by the basic waveforms, you can become familiar with the palette and begin to understand how the basic elements interact.

Only once you have a knowledge of the broader tones that are available and how the elements interact will you be able to call up the sounds you need when designing a patch rather than hoping for a happy coincidence. You will understand how to create the sound you need for your track and you will also find that programming becomes much faster.

Another aspect of listening to all of these sounds is that we have ruled out some options. For instance, detuning an oscillator when the other oscillator is several octaves higher does not necessarily give a useful tone (unless you are trying to create a special effect). Now we know this information, we don't need to spend time experimenting.

With the combinations we have looked at so far, there have only been two waves. When you design sounds you can of course use more waves, for instance you may want to use a combination of two slightly detuned sawtooth waves with a square wave dropped an octave lower to fatten up the sound.

You should also remember that with modern synthesizers offering many oscillator slots, you don't have to use all of the slots when the key is struck. If you want you could allow some elements of the sound to fade in gently over time.

Combining Waves—Practical Patches

If you have worked your way through the examples, you've probably had enough of listening to detailed nuances for a lifetime. Let's move on and create a few practical patches in Z3TA+ using some of the new tones we've created by combining waves.

The following group of sounds has been built with some of the wave combinations we have just listened to. You will also see that I have used some of the modulation options which are discussed in the next chapter.

Lead Saw Squ +1 Filter

For this first patch I took the patch we looked at earlier called Saw Squ +1 Filter. The original patch didn't necessarily appear to be most immediately useful, however it did have a tone that would cut through a mix without totally dominating, making the wave combination useful as the basis for a lead sound.

To make the patch, I made the following changes to the earlier patch:

69

- First I reduced the polyphony (Poly) to one and engaged portamento (by setting Porta Mode to Fing, Var, and the glide time, that is P.Time, to about 0.4 seconds).

- I set the Bend Up range to four semitones (a major third) and Bend Down to -12 semitones (an octave).

- To make the filter a bit more interesting I increased the filter Resonance to around 12 dB—to balance this I set the cut-off slope (the filter Type) to 36 dB/octave.

- To complete the patch I made it velocity responsive in two places (both set through the modulation matrix):

    o   filter cut-off (see below), and

    o   filter level, so the volume of the patch is controlled by the filter's level which is controlled by key velocity.

The next two chapters will explain more about modulation. Let me explain how the filter in this patch operates—you can figure out why it works once you have read the next chapter.

The filter cut-off reacts to both velocity and the modulation wheel. For normal playing I suggest you set the modulation wheel to its minimum—this will allow you to control the cut-off frequency by using velocity alone. However, this is not the only control you have—you can control the filter in real time with the modulation wheel, so when a note is sustaining you can close the filter by pushing up the wheel and then reopen it as the note sustains.

The velocity controlled filter cut-off combined with the high resonance gives a wide range of playable tones. The modulation wheel then allows for far more extreme (and potentially unnatural control over the filter cut-off).

Harpsi Two Saws +2

This is a synthetic harpsichord based on Two Saw +2 Filter Octaves. I chose this combination because of the way that it interacts with the filter. I could have used other combinations (such as Two Saw Phase 90 for different tonal variations).

This patch is unrealistic when compared to a real harpsichord in two areas:

- first, its sound only bears a passing resemblance to a harpsichord (in that it is bright and quite staccato), and

- second, this synthetic patch is touch sensitive.

The first change from the basic patch is that the filter cut-off is now controlled by Envelope One (and the filter has the secondary effect of controlling the volume of the patch—when the envelope reaches the zero level the filter is closed and so no note is audible).

The second change is that the Level of oscillator two is modulated by velocity. This means that at quieter volumes the oscillator cannot be heard. At higher velocities the oscillator can be heard as an added brightness in the attack of the note. With the operation of the filter the second oscillator is only heard during the attack phase of the note—once the filter envelope has reached its sustain phase, the second oscillator cannot be heard.

Organ Two Squ +3 Filter

This patch is based on Two Squ +3 Octave Filter and takes advantage of the organ like quality of that earlier patch.

From the base patch I made two main changes, the first to the filter and the second to add some vibrato. I will deal with each in turn.

For the filter, I wanted to do two things:

- add an element of velocity sensitivity so that the tone can be controlled by the player, and

- set the filter so that it closes as notes get higher: this will ensure that the tone is more consistent across the keyboard.

I wanted to add some vibrato to simulate the effect of a rotating speaker. To do this I set two local LFOs (in other words polyphonic LFOs—see Chapter 5: Modulation and Other Ways of Messing with Things) to modulate the two oscillators (one LFO per oscillator) and set the frequency of the LFOs to be slightly different. The amount of modulation was set to be the minimum that could be heard.

Other Sound Sources

Most of the sound sources we have considered so far are likely to be found in conventional subtractive synthesizers. However, there are many other sound sources (using this term in the widest possible sense). Some of these other sound sources are a combination of existing sounds and some are sounds created in wholly new ways.

Samples

All of the synthesizers featured in this book (apart from Vanguard) have the ability to use samples as wave sources. Surge and Z3TA+ allow only single cycle waves to be imported whereas Rhino and Wusikstation allow full length multi-samples to be used. Cameleon 5000 resynthesizes sounds which gives other sonic option.

The ability to include other waveforms effectively makes the sonic options limitless.

The book does not cover sampling. However, if you do want to learn more about sampling, then I recommend you check out "Sample this!!" written by me and Klaus P. Rausch.

Frequency Modulation (FM)

Highly complex waveforms can be created by taking two waves and using one to modulate the frequency of the other.

FM has developed considerably since the earliest FM synthesizers—different wave forms can now be used and FM synthesizers have filters to give even more sonic options. FM is described in greater detail in Chapter 7: Frequency Modulation Synthesis and Chapter 7 Supplement: Frequency Modulation Synthesis with Surge.

There are other forms of modulation which are often used. For instance

- Phase Modulation which is similar to Frequency Modulation. Z3TA+ offers this option—its tone is more restrained and controllable than FM.

- Ring Modulation which is the effect of combining the sum and difference between two waveforms' frequencies. As I will demonstrate in a moment, the results can be unpredictable.

Feedback

I have already mentioned feedback in the last chapter. This process is available in Surge and simulates the effect of feeding the audio output back into the audio chain. In highly simplistic terms, it works much like adding a second oscillator which has its phase very closely linked to the first (although you are feeding back a signal which has been through filters and envelopes etc). The two factors which have the greatest influence on the effect of the feedback are:

- the level of feedback, and

- the polarity—when inverted it works much like an inverted waveform to cause sound cancellation.

Wave-Sequencing

In basic terms, wave-sequencing is stepping through different waveforms in a predetermined order. You could have a sequence which plays a sawtooth wave for a beat, then a square wave for a beat, a sine wave for a beat and finally a triangle wave for a beat before looping back to the start of the sequence.

However, simply cycling through four or five basic waveforms is unlikely to give a sonically rich results. Hence most synthesizers that are capable of wave-sequencing (such as Wusikstation, and to an extent, Surge) contain a wide variety of waveforms to allow more possibilities and greater sonic nuances. The waves can then be stepped through to give rhythmic effects or cross-faded between to give constantly shifting sounds.

Wave sequencing is described in greater detail in Chapter 8: Wave-Sequencing.

Additive

Additive synthesis often sounds simple in theory—you add some sine waves together to create a new waveform. It is easy to pick a few sine waves and create a new waveform. However, the difficulty comes in choosing the right sine waves and then controlling them over time.

Additive is described in greater detail in Chapter 9: Additive Synthesis.

Further Sonic Examples

I now want to illustrate some of these other sonic techniques that I have just mentioned. For this group of examples I am going to use Surge.

01 Feedback Shifter

This first patch give an example of the effect of feedback.

The patch is a simple sawtooth waveform without filtering. The audio signal is then fed-back into the start of the audio chain. If you strike a key, and hold it, you will hear a varying tone. This is because the amount of feedback, and its polarity, is being controlled by a low frequency oscillator.

If you hold the key the sound will shift from the raw waveform and the amount of feedback will then increase. Having reached 100% feedback, the feedback will start to decrease, passing through the zero feedback point from where reversed polarity feedback will start to be added. When 100% inverse polarity feedback is reached, the amount of feedback will start to reduce until it reaches 0% when the process will start again.

As you can hear, in a situation like this, the amount of feedback has more of an effect on the volume of a note than it does on tone. However, with more complex patches the effect can be more noticeable. Go and play with this control and find out more.

02 Square Saw Shifter

This second patch in this group is intended to illustrate the constantly variable tone in Surge's Classic oscillator. There are three basic sounds which can be generated by this oscillator:

- a square wave

- a sawtooth wave

- two sawtooth waves 180 degrees out of phase (which as we have seen earlier result in a sawtooth wave which has been raised by an octave).

These are not simply three choices, instead the oscillator allows you to constantly shift between the square wave at one extreme, the sawtooth in the middle, and the octave sawtooth at the other extreme. In between the extremes you get a mix of square and sawtooth, or sawtooth and changing phase sawtooth. In the middle, there is pure sawtooth, of course.

If you listen to this patch and hold a key, you will hear the sound shifting between these three points. In practice you are unlikely to use a shifting wave shape like I have done here. However, this patch does illustrate a small element of the vast tone palette that is available in the apparently basic waveforms in Surge.

03 Waveshape Shifter

With 03 Waveshape Shifter, you will hear a plain sawtooth wave being transformed by the Waveshaper using the Sinus option. This is one of the more aggressive transformations.

In this patch, the amount of wave shaping starts at zero and then is progressively applied positively. When it reaches the maximum amount, the amount reduces, passing through zero until it reaches 100% negative wave shaping. After this it progressively returns to zero whereupon the process starts again.

As you can hear, the Waveshaper offers a huge range of additional tones and textures. In a musical context you will probably find this control is most useful when used with discretion and subtlety. However, I do encourage you to experiment, both with the amount of wave-shaping and with the different Waveshaper options.

04 Aggressive Wave

With 04 Aggressive Wave, I have taken a sawtooth wave and applied three transformations:

- first, the wave shape has been changed slightly by pushing the Shape slider increase the prominence of the out of phase sawtooth wave

- second, some (positive) feedback has been added (around 50%), and

- third, some Asymmetrical Waveshaping has been added.

The result of these three transformations is that a plain sawtooth wave has been given a much sharper, aggressive tone. If you don't like it, please feel free to keep fiddling with the controls until you find something you like, and then add some shaping.

05 Ring Modulation +7

05 Ring Modulation is the first example ring modulation patch in this book. It has been created with two sine waves—the second is pitched seven semitones above the first. Together these sine waves sound quite dull. However, when their ring-modulated output is also added (by un-muting the 1×2 fader), the sound becomes more interesting. If you mute the oscillator 1 and oscillator 2 faders, you can hear the ring modulation on its own.

06 Ring Modulation +9

I chose the first ring modulation example because I knew the waves would work together. With this second example, 06 Ring Modulation +9, you will hear a considerable dissonance within the sound. Mute the direct output from the two oscillators and you will hear the ring modulation on its own. As you can hear, this is a much cleaner sound and is not unlike the ring modulation (without the oscillators) in the previous example.

Experiment and you will find that the results you get are very much dependent on the waves you choose, and the interval between the waves.

07 RM +7 Organ-Like

This third sound, 07 RM +7 Organ-Like, gives an example of ring modulation in practice. As you might expect, it is based on the first ring modulation example, 06 Ring Modulation +7.

The key difference is that the level of the second oscillator is modulated by a low frequency oscillator so its volume varies giving a Leslie-esque rotary sound to this organ-type patch.

Chapter 5
Modulation and Other Ways of Messing with Things

So far in this book we've looked in turn at each of the main element that are pulled together when constructing a sound—the oscillators, the envelopes and the filters. We're now going to move on and look at *how* these elements fit together.

Modulation goes to the heart of creating sounds. It is the technique by which a synthesizer creates the varied nuances of sound. Without modulation sounds may be rich, thick, fat, bright or delicate. However modulation will take a basic sound and animate it, giving it life in the hands of a skilled player.

Modulation: The Concept

The concept of modulation in a synthesizer is simple—something is changed by something else and usually in real time.

Conventionally, a "source" will modulate a "destination". So if we think about vibrato, a low frequency oscillator would act as the source and the pitch would be the destination.

Modulation in the Real World

Before we get to the detail, let's think about an acoustic instrument such as the piano. Taking a simplistic view, the factors that have an effect on the sound are:

■ velocity

■ pitch, and

■ time.

Effect of Velocity on a Sound

Velocity is another way to describe loudness, which in the case of a keyboard would be determined by how hard the key is hit. Classically trained musicians are used to seeing markings to indicate pianissimo (very soft) ppp, fortissimo (very loud) fff and so on. With MIDI we have 127 levels of loudness—this range is called "velocity". So instead of fff you may have a note with velocity in the range of 113 to 127.

With a piano, the velocity (the loudness) of a note has an effect in three main areas:

- the volume—there is a direct link between the velocity and how hard a note is struck

- the tone—the harder the key is hit, the brighter the tone, although the range of tones between the softest and the loudest notes may not be as dramatic as the volume differences

- the sustain time of the note—the louder the note, the longer it sustains.

Effect of Key Pitch on a Sound

Clearly the piano keyboard controls the pitch of the note played. However, there are other factors affected by the pitch:

- the tone—the higher the note, the brighter it sounds

- the sustain time of the note—the higher the note, the less time that it sustains.

Effect of Time on a Sound

The length of time that a note is held is also a factor in the tone of a piano. Over time:

- the sound will get duller (however bright it may have been initially), and

- the volume will decay until it finally reaches zero.

Combining the Effect of Velocity, Pitch, and Time

A piano note is far more complex than may be suggested here, but in broad terms you can see that:

- The initial volume of the note is affected by one factor only—velocity.

- Tone is affected by three factors—the initial velocity of the note, the pitch of the note and the length of time it is held. However, you should also note that:
  - the tone *increases* in line with the increase in pitch (so the higher the note, the brighter it becomes), but
  - the sustain time *decreases* in line with the increase in the pitch (so the higher the note, the less time that it sounds for).

- Sustain time (and also volume over time) is affected by two factors—the starting volume (which is in turn controlled by the initial velocity), and the pitch.

To mimic some of the characteristics of an acoustic instrument, we're going to need to use modulation to control the synthesizer.

Modulation Destinations

We will approach modulation in a slightly backwards way and look at modulation destinations before we look at modulation sources. Once we understand what can be modulated, we can work back to choose the best source to create the modulation.

I should point out that the following list of modulation destinations is, of course, not complete. The synthesizers covered by this book have more features that can be modulated: this list just covers the main destinations that you are likely to want to control.

Volume

Volume is likely to be one of the main destinations you modulate. You may want to:

■ control the volume of your whole patch—for instance to make the patch touch sensitive or to add a tremolo effect, or

■ control the volume of individual oscillators so you can cross–fade between two sounds.

Filter

The two main modulation destinations in the filter block are:

■ the cut-off frequency, and

■ the resonance.

Often musicians will want to control both in real time—the cut-off frequency to make a sound brighter or duller and the resonance to add some sharpness and bite to the sound.

Some synthesizers (such as Z3TA+) also allow you to control the separation of the filters. This is where the filters are a combination of several other filters—with the separation control, the cut-off frequencies of the stacked filters can be addressed individually so that one will be raised in relation to the other (or others) giving differing resonant peaks and a different tone to the filter.

Pitch

The effects that can be obtained by modulating pitch include vibrato and trills as well as weird special effects. Subtle pitch modulation is also important when trying to mimic the properties of natural instruments—for instance a slight and very short rise in pitch at the start of a note will often help recreate part of the attack of a brass instrument or a plucked string instruments.

Vanguard makes a distinction between Pitch and Detune, and Wusikstation makes a distinction between Pitch and Fine—in essence there is no difference between these as destinations, but Detune/Fine does allow a greater level of detail in the control.

77

When using an LFO to introduce vibrato you will probably have to apply less vibrato than a player would apply to a natural instrument. LFO vibrato usually has a fixed frequency (but see below) and so the effect will quickly sound mechanical. Reducing the depth of the effect helps to stop some of the mechanical feel becoming too prominent.

Vibrato Speed

When a musician applies vibrato to a real instrument, both the speed and the depth of the vibrato increase—it therefore make sense that LFO speed should be a modulation destination (and the depth of the effect should also be controllable).

Pulse-Width

A fast way to change the tone of an oscillator or to fatten up a sound is to play with the pulse-width modulation—this was discussed in greater detail in Chapter 4: Sound Sources.

Pan

The classic use for panning is auto-pan type effects where the sound moves regularly between the left and right channels. There are however more subtle uses of panning, for instance panning can be related to the pitch of a note with low notes panned towards the left and higher notes towards the right as if you were sitting in front of a piano.

FX

FX sends and returns can be modulated. Sends can be controlled so that FX is added for emphasis on certain notes. Returns can be controlled so that, for instance, the FX signal is suppressed while the signal that is being fed to the FX is present—this "ducking" effect means that a patch will sound less muddy when it is played since the FX won't swamp the sound.

If you want to get more complicated, you may want to change controls in real time—for instance increasing chorus depth as vibrato increases.

You can also modulate other effects such as EQ or delays.

Wavetable Position

Surge offers a range of wave tables. Instead of single wave source, these allow you to step (or sweep) through a range of different waves in order to create a shifting tone. You can also set up this process manually with Wusikstation.

I will look at wave-sequencing in greater detail in Chapter 8: Wave-Sequencing.

Sample Start Point

Wusikstation allows you to modulate the sample start point. You may use this to cut the start of a sample, which may be useful if either:

■   you want a sharper start—perhaps the sample fades in, or

- you want to avoid a sound at the start of the sample, for instance if you want to bypass a guitar string being picked but keep the sound of the string sustaining.

Envelope Speed

Key tracking was mentioned earlier—the ability to control envelope speed is useful, even for entirely synthesized sounds, to give the sound a more natural quality.

Modulation Sources

There are many modulation sources within synthesizers—the most common are LFOs and envelopes. It is also possible for external sources to act as modulators—this is the case with velocity, the pitch bend wheel and the modulation wheel which are all dependent on the musician. However, player expression controllers can be used either as modulation sources or as modulation source controllers. For instance:

- The modulation wheel could be used to control the filter cut-off—in this case, the modulation wheel would be acting as a modulation source.

- Equally, the modulation wheel could be used to control the amount of vibrato applied to a note by an LFO—in this case the modulation wheel would be acting as the modulation source controller.

The main purpose of player expression controllers is, as the name would suggest, to allow the musician greater control over the notes they play. The earliest synthesizers essentially gave the option of note on and note off and little more—today there are more options, although in practice the main controllers that are used are velocity, the pitch bend wheel and the modulation wheel. However, the choice of controllers used is usually quite specific to individual musicians.

The main modulation sources are summarized below—LFOs are explained separately in the following section.

Velocity

Velocity is often used to control:

- the loudness of a note

- the tone of a note by modulating the filter cut-off, and

- the volume envelope attack time (although from the six synthesizers considered in this book, only Cameleon 5000, Rhino, and Surge allow you this level of control).

Velocity can either be applied directly to modulate the destination, or it can be used as a modulation source controller to control another source (for instance, an envelope) which is modulating the destination. When velocity is used as a source controller, you can often achieve smoother results.

Velocity Layering

With Wusikstation and Rhino there is an additional aspect to consider: velocity layering. This is the process where multi-samples are constructed with different samples assigned to different velocity ranges, so for instance you could layer a soft, a medium, and a loud piano sample. This will give a different source tone for the patch at different velocity levels. Velocity layers can be used in conjunction with velocity being used as a modulation source so, for instance, you could use a delicate amount of filter modulation to hide the steps between your three sample layers.

Envelopes

There are many things you can do with envelopes. If you apply the envelope to the filter negatively it will close down the filter rather than open it up. Starting a note with the filter closed and then opening it quickly may be another useful technique if you are trying to simulate the "spit" of a brass instrument. For further simulation, you might also want to add a pitch changing envelope at the start of a brass note, too.

Envelopes can act as modulation sources or can work as modulation controllers—a typical use as a modulation controller would be to slowly fade in and then fade out some LFO vibrato. You will see that Surge has a dedicated envelope to control its LFOs.

Pitch Bender

The pitch bender is often hardwired to the pitch bend function so you won't be able to assign it to control volume (as an example). That being said, some synths allow the wheel to be assigned to any control.

Modulation Wheel

Conventionally the modulation wheel is wired to the vibrato depth and so acts as a modulation controller.

However, it doesn't have to be so—the modulation wheel can also be used as a modulation source (for instance, Vanguard's modulation wheel is hardwired to the filter cut-off).

Key Tracking

Key tracking is the process where changes are made according to the pitch of a note.

Take an example, if you pluck the bottom E string of an acoustic guitar, the sound will last for over 10 seconds. However, if you play the top E string at the highest fret, the sound will barely last for 3 seconds. With key tracking we can seamlessly control the length of an envelope to reflect the behavior of the acoustic instrument.

Key tracking can be applied in many ways—for instance, the cut-off point of a filter can be controlled by key tracking so that higher notes are brighter than lower notes.

XY Controllers

Both Wusikstation and Z3TA+ have XY pads (see Figures 5.1 and 5.2). These allow you to control two destinations from one controller—for instance, you could control a filter's cut-off frequency and the filter's resonance simultaneously. Alternatively you could mix the output of four oscillators.

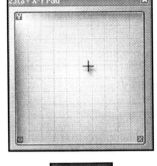

Figure 5.1 Z3TA+'s XY controller.

Figure 5.2 The Wusikstation XY controller.

XY pads can work as a modulation source or as a modulation source controller.

Cameleon 5000 gives similar functionality with its morph square (see Figure 5.3). However, the morph square offers additional functionality in that it can be controlled by dedicated multi-part envelopes.

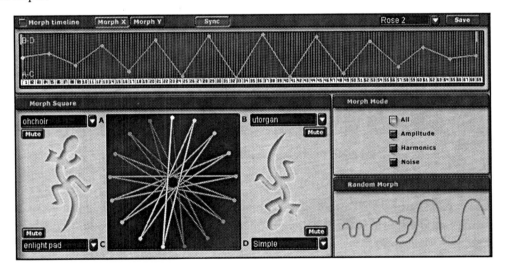

Figure 5.3 Cameleon 5000's morph square with (one of two) controlling envelopes.

81

Aftertouch

If your keyboard (or other MIDI controller) gives aftertouch messages, then the pressure on the keyboard while notes are sustained can be used to control your patch. You could, for instance, use aftertouch to control volume to give a swell to a held string chord.

Aftertouch can either be applied directly to modulate the destination, or it can be used as a modulation source controller to control another source which is modulating the destination.

Expression Pedal

Often the expression controller is hardwired to a synthesizer's volume. As with the modulation wheel, it can also be used as a modulation controller.

Wusikstation Wave-Sequence Layers

Wusikstation has two wave layers, each of which has 12 sequence lanes which are conventionally used for wave-sequencing. However, these can also be used as a modulation source (or a modulation source controller)—for instance, if a lane is assigned to control the volume as shown in Figure 5.4, then gating type effects can be generated easily.

Figure 5.4　Using Wusikstation's controller lane to create gating effects.

Surge Step Generator

As part of its LFO section, Surge has a step generator (see Figure 5.5). This allows control of up to 16 steps. For each individual step you can set the and so create rhythmic modulation effects.

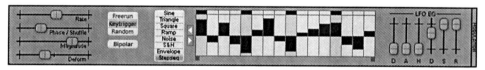

Figure 5.5　The Step Generator in Surge's LFO.

Oscillator Output

Rhino, Wusikstation and Z3TA+ all allow the output of an oscillator to be used as a modulation source. The most frequent use for this would be for creating FM sounds. Surge builds on this principle with some dedicated FM algorithms which I will discuss further in Chapter 7 Supplement: Frequency Modulation Synthesis with Surge.

Random Generators

Both Wusikstation and Z3TA+ allow a random generator to be included as a modulation source. This can often help to remove the mechanical quality of some sounds.

MIDI Control Code

Lastly, MIDI control code can be used as a source or as a source controller.

LFOs as a Modulation Source

A whole book could be written about LFOs—especially with all of the options that are available under Z3TA+.

The low frequency oscillator is an oscillator which oscillates at a low frequency—for our purposes it usually oscillates below the audio threshold. As you know, in addition to working as a vibrato source, the LFO can also modulate:

- volume to give tremolo effects, and
- the filter cut-off to give wah-wah or gating/stuttering type effects—especially when a square or pulse wave is selected as the modulation source.

You can, of course, modulate many other destinations with an LFO.

Main LFO Controls

The main controls of an LFO are wave shape and depth.

Wave Shape

Most synthesizers offer several LFO shapes:

- Vanguard only offers one (but that's all it needs).
- Cameleon 5000 offers three.
- Surge and Wusikstation offer six (although Surge also has the Step Generator and a dedicated LFO envelope).
- Rhino and Z3TA+ both effectively have limitless options.

The most common LFO shapes are:

- **Sine.** The sine wave produces an even and rounded modulation source.
- **Triangle.** The triangle wave has virtually the same effect as the sine wave, but the changes are constant. This may give a crisper sound when compared to the sine wave. Both the sine and triangle are ideal waves for vibrato, tremolo or wah-wah effects.
- **Sawtooth.** Depending on whether the wave is applied positively or negatively, the sawtooth provides either

- o   a rising modulation source which drops abruptly after reaching the maximum value, or

- o   a falling modulation source, rising to a peak and then gently falling.

The sawtooth wave is often used for modulating the filter cut-off.

- ■   **Square.** The square wave is either at its maximum or its minimum. This wave tends to be used as an LFO to give rhythmic effects. It can also be used to modulate the pitch to give trill type effects.

- ■   **Random.** The output from the LFO is a random value. Again this wave is often used for creating rhythmic effects.

- ■   **Noise.** Noise is a less commonly used LFO source.

If you look at Rhino and Z3TA+ you will see that you have many more LFO shape options—these are mentioned at the end of this section.

Speed

The speed control affects the frequency of the LFO. If an LFO is used as a vibrato source, then the speed control will affect the frequency of the vibrato. Most LFOs have a range of around 0.1 cycles per second to about 20 cycles per second.

The LFO speed can be synchronized with the tempo of a track so that the time it takes the LFO to complete a whole cycle is a subdivision of the track's tempo. This option is often used in conjunction with rhythmic effects.

Depth

The depth of an LFO's modulation (that is how much it affects the modulation destination) needs to be carefully balanced so that an appropriate range of its effect is controlled. For instance, if you are using an LFO to give some vibrato, you probably want a range which is less than (plus or minus) one semitone. With Z3TA+ it is easy to select the range by using an appropriate curve, for the other synthesizers you will have to judge the range by ear.

Phase

LFO phase may be important. If the LFO phase is synchronized to the start of a note then when the note is triggered the effect of the LFO will be known. By contrast, if the LFO is running freely it may be at any part of its cycle when the note is triggered (this random element may be desirable for ensemble type effects).

Z3TA+ also allows you to control where the LFO starts its cycle, so for instance you could select a triangular wave and direct it to start at the top of its cycle (where the wave will have maximum effect immediately).

Monophonic/Polyphonic LFOs

LFOs can be monophonic or polyphonic. If you are using Z3TA+, monophonic = global, and polyphonic = local. In Surge, the LFOs are polyphonic, but Scene LFOs (SLFOs) are monophonic.

Monophonic (Global) LFOs

In Wusikstation and Z3TA+ (LFOs 1 to 4) the LFOs are monophonic.

With monophonic LFOs, a single oscillator is used whenever that LFO is invoked. So if you are using an LFO to add some vibrato to a pad, when you play a chord all of the notes will have the rise and fall of their pitch synchronized. The effect may not be particularly natural in this instance.

Polyphonic (Local) LFOs

In Cameleon 5000, Rhino, Surge, Vanguard, and Z3TA+ (LFOs 5 and 6) there are polyphonic LFOs. With a polyphonic LFO, each separate note has its own LFO which will start its cycle at a different time from any other LFO.

Using our pad example again, this could mean that if you are using a polyphonic LFO to modulate a chord, some notes could have their pitch raised while others could have their pitch flattened. This may give a more desirable result for pad, but is less use for rhythmic sounds.

Rhino LFOs

Rhino gives you two options for an LFO source:

■ you can use any oscillator as a wave which means you have a limitless LFO shapes available as you can import your own waves, or

■ you can draw your own modulation envelope as shown in Figure 5.6 (noting, of course, that there are many LFO envelopes that come with Rhino). The deployment of the LFO envelope is discussed in the next chapter.

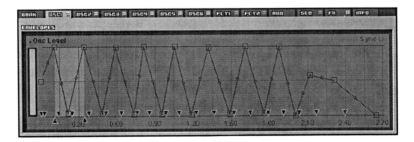

Figure 5.6 An LFO envelope drawn in Rhino.

Z3TA+ LFOs

As with Rhino, Z3TA+ gives you a wide range of waveforms to choose from, including the option to import user waves. Unlike Rhino, Z3TA+ has a bank of six dedicated LFOs (see Figure 5.7) with a host of very powerful features.

Figure 5.7 The Z3TA+ LFO block.

In addition to the wave selector and the speed slider, you can also sync an LFO to the track's tempo and control the phase of an LFO. So far, so good. But Z3TA+ can do more.

Delay and Fade

- Delay sets the time between the note being struck and the LFO starting its cycle.

- Fade sets the time (after the delay phase has been completed) over which the LFO fades in—the LFO reaches its full value at the end of the fade phase.

If the LFO is controlled in the control column of the matrix (more later) then the Delay and Fade times are not affected by whether the controller allows the LFO to have effect.

Morphing

The Z3TA+ LFO isn't simply one LFO: there are two waveforms (both running at the same frequency). Z3TA+ gives you the opportunity to

- combine these two waves in various ways

- morph between the waves, and

- join part of one wave with the other.

Lastly there is the "one-shot" option so instead of working as a repeating cycle the LFO will complete one cycle and then cease to have effect. This may be useful if you want some pitch bend at the start of a note.

Surge LFOs

As well as offering the Step Generator, Surge has one other great trick up its sleeve: a dedicated envelope controller attached to each LFO. This is a DAHDSR envelope (delay, attack, hold, decay, sustain, release envelope) which is described in Chapter 2: Envelopes. See Figure 2.6 for an illustration of the envelope and Figure 5.8 to see the envelope controlling a sine wave LFO.

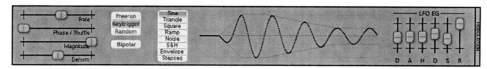

Figure 5.8 The Surge LFO envelope controlling a sine wave LFO.

Adding an envelope to the LFO allows a huge amount of control over the LFO, going beyond the Delay and Fade options offered by Z3TA+. As you would expect, Surge also allows its LFOs to be synchronized to the track's tempo and for control over the phase of its LFOs.

Chapter 6
Modulation in Practice

Modulation is implemented in different ways in different synthesizers. For some there is a dedicated control for any available modulation, for others there is a matrix. In this chapter I am going to look at some of the featured synthesizers in a bit more detail to see how they implement modulation. First I will look at the functions of the synthesizers—later I will consider how to use these controls in a practical situation.

Modulation in Vanguard

Vanguard offers a range of dedicated modulation controls—this means that each modulation source and the modulation destination are fixed. The advantage of this design is speed when using the synthesizer—the potential disadvantage is a lack of flexibility (although you would be hard pressed to find a modulation option that isn't available as Figure 6.1 shows).

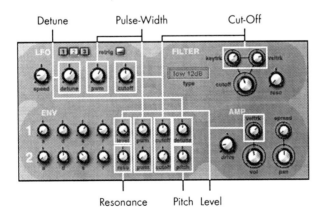

Figure 6.1   The modulation options in Vanguard.

Vanguard allows you modulation control over:

- pitch

- pulse-width

- filter cut-off

- levels, and

- resonance.

These modulation destinations can be controlled by these sources:

- velocity

- envelopes

- LFOs

- Modulation wheel, and

- key position

As table 6.1 shows, not every source can modulate every destination.

Table 6.1 The available modulation sources and destinations in Vanguard.

	Pitch	Pulse-Width	Filter Cut-Off	Filter Resonance	Volume
Velocity	✗	✗	✓	✗	✓
Envelopes	✓	✓	✓	✓	✓
LFO	✓	✓	✓	✗	✗
Modulation Wheel	✗	✗	✓	✗	✗
Key Position	✓	✗	✓	✗	✗

Let's look at the options in more detail.

Modulating Volume in Vanguard

- Volume can be controlled over time by using Envelope One. The volume envelope will control the volume of all of the oscillators and the depth of the modulation is controlled by the Level knob in the Envelope One lane. For negative values (when the Level knob is turned to the left) the effect of the envelope will be to cut the volume.

■ The second way to control the volume is by velocity. The Veltrk knob in the amplifier section controls the effect that velocity has on the volume of the patch—turned to the right and the keyboard has to be hit harder to increase the volume. If the knob is turned to the left, then higher velocities cause the volume to be cut.

Modulating Pitch in Vanguard

■ Each oscillator has a dedicated LFO to apply vibrato separately to each corresponding oscillator. The speed of the vibrato is controlled by the Speed knob and the depth of the vibrato (that is how far from the centre frequency the note is wobbled) is controlled by the Detune knob (both knobs in the LFO bank).

■ Envelope One can be set to modulate the pitch (plus or minus one semitone) of all oscillators simultaneously—this is controlled by the Detune knob in the Envelope One lane.

■ Envelope Two can be set to modulate the pitch (plus or minus 4 octaves) of all oscillators simultaneously—this is controlled by the pitch knob in the envelope two lane.

Modulating Filter Cut-Off in Vanguard

There are eight controls that affect how the filter cut-off is modulated (and there is the Cut-Off knob itself).

■ Both Envelope One and Envelope Two can modulate the filter cut-off. The effect of the envelope is controlled by the Cut-Off knob in each envelope's lane. The envelope can be applied to open the filter (if the Cut-Off knob is turned to the right) or to close the filter (if the knob is turned to the left).

■ Each LFO can be applied to modulate the filter (for instance, to give a wah-wah effect). The effect is controlled by the Cut-Off knob in the LFO bank—turned to the right and the LFO opens the filter in relation to the cut-off frequency; turn the knob to the left and the LFO works to close the filter in relation to the cut-off frequency.

■ The Keytrk knob in the filter bank determines how the pitch of a note affects the filter cut-off frequency. When the knob is turned to the right, the filter opens up as higher notes are played. As the knob is turned to the left, the filter closes as higher notes are played, giving brighter bass sounds and muffled treble sounds.

■ The Veltrk knob in the filter bank determines how the velocity of a note affects the filter—this knob works in the same way as the Veltrk knob in the amplifier section. As there are separate Veltrk knobs in each section, a greater degree of control can be exercised over the volume and filter modulation. As with the amplifier section, turn the knob to the right and the keyboard has to be hit harder to open the filter. If the knob is turned to the left, then higher velocities will cause the filter to be closed.

■ Finally, the modulation wheel is hardwired to the filter cut-off and will open the filter.

Modulating Resonance in Vanguard

There is only one modulation source for resonance—Envelope Two. The effect that the envelope has on the resonance is controlled by the Reso knob in the Envelope Two lane. Turned to the right and the envelope will increase the resonance, but turned to the left and envelope two will decrease the resonance.

Modulating Pulse-Width in Vanguard

There are several options for modulating the pulse-width of the oscillators. For any of the controls to have an effect, a waveform must be selected that is capable of having its pulse-width modulated—these waves are indicated by PWM in their name.

- Each LFO can be applied to modulate the pulse-width of its respective oscillator. The effect is controlled by the PWM knob in the LFO bank.

- Both envelopes can modulate the pulse-width of all of the oscillators. The effect of the envelope is controlled by the PWM knob in each envelope lane.

Working with Vanguard's modulation structure

Vanguard is designed to be simple (and therefore fast) to operate. It succeeds in this aim and therefore cannot be criticized for not having some of the more complicated features of the other synthesizers in this book.

Only having two envelopes may be seen as limiting, but in practice, this is unlikely to be an issue. Often envelopes are used in conjunction with velocity controls. As the velocity based modulation destinations are separately controlled, there is sufficient control.

The number of envelopes may also have been a problem if each of the oscillators could be separately controlled—the oscillators cannot be separately controlled and so there is no necessity for one envelope per oscillator.

Three separate, polyphonic LFOs give considerable flexibility. If LFOs are applied to more than one oscillator, it is a fiddly task to precisely replicate the LFO settings. This means that you are more likely to get the settings close but not perfect—this lack of perfection tends to add more movement to a patch.

It is unfortunate that vibrato depth (as an example) cannot be controlled by the modulation wheel—as part of the simplicity of Vanguard, the modulation wheel is hardwired to the filter cutoff.

Modulation in Rhino

Rhino takes a unique approach to modulation, which is not surprising for a synthesizer that uses modulation (in Rhino's case, frequency modulation) to produce sounds.

The modulation for each oscillator and each filter is individually set on that oscillator or that filter's page, so the effect of velocity, key tracking, and envelopes etc are set individually for each oscillator. The only exception to this is where an oscillator is modulating the frequency of another oscillator (whether in the normal FM manner or acting as an LFO). Where the oscillator is the modulation source, the routing and the depth of the modulation is determined in a routing matrix.

A practical example may help to explain how the modulation system works in Rhino.

Using Modulation in Rhino

For this example we will take a simple patch, Modulation Example, which has one sawtooth oscillator running through a filter. Velocity will affect the patch in two ways:

- first, it will make the sound brighter as a key is struck harder, and

- second, it will change the attack time of the volume envelope—at lower velocities the sound will fade in slowly while at higher velocities, the attack will be much faster.

Modulating Attack Time

In Rhino, each oscillator's level envelope is always applied with 100% depth. If you look at the oscillator's level envelope in Figure 6.2, you will see two time markers have been indicated:

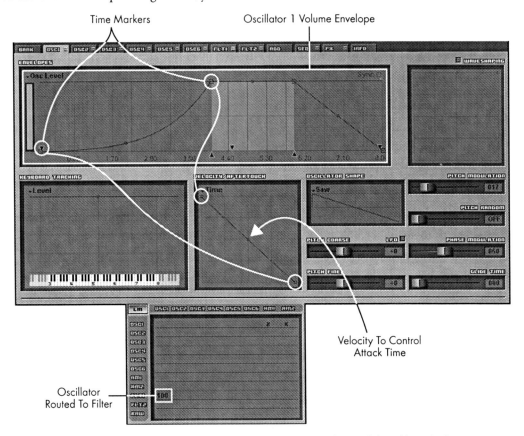

Figure 6.2   Setting up Rhino so that attack time can be modulated by velocity.

91

- The most obvious marker is the breakpoint at the end of the attack phase. If there is no modulation of the attack time, this is the attack time that you will hear. For this patch, the default attack time is fairly slow.

- The other time marker is the circled downward arrowhead—this indicates the attack time at maximum modulation. You will see that at maximum modulation, the attack time is quite fast.

Next look at the Velocity/Aftertouch window. You can see a diagonal line—this controls how the modulation attack time (in that window's drop-down menu, Time has been chosen) is affected by velocity. The grid works as follows:

- velocity values count from left to right—from 0 (on the left) to 127 (on the right)

- the top axis represents no change (so the time marker in the envelope takes its full value)

- the bottom axis represents the downward arrow taking its full value.

So if you set the line horizontally at the top of the square, velocity will have no effect—the attack time will always be that set by the envelope. If you set the line horizontally at the bottom of the square, velocity will have no effect—the attack time will be that set by the downward arrowhead. Indeed, whenever the line is horizontal, velocity will have no effect.

In this example patch, the line goes diagonally downwards. This means:

- at lower velocities, the attack time set by the envelope predominates (and so the sound fades in slowly)

- at higher velocities, the attack time set by the downwards arrowhead predominates (and so the attack is much faster), and

- at velocities between the two extremes, the attack time varies depending on the velocity.

If the line were to go diagonally upwards, then the behavior of the modulation would be reversed (so the attack time would be faster at lower velocities and slower at higher velocities).

You will notice that in this patch velocity does not affect the volume of the oscillator's level (although this would be easy to achieve using the Velocity/Aftertouch window and choosing Level).

Modulating the Filter Cut-Off

There are four factors controlling the filter cut-off (see Figure 6.3):

- First (and most obviously), the cut-off slider in the main filter control is set to 16 (on a scale that goes to 100).

- The second factor controlling the filter cut-off is the filter envelope depth slider (labeled Cut-off Env Mod). This works in a slightly different way to the way that this type of control works on other synths in that it allows the envelope to effectively cut the filter. Unlike other synthe-

Filter Cut-Off Envelope   Filter Envelope Depth Control

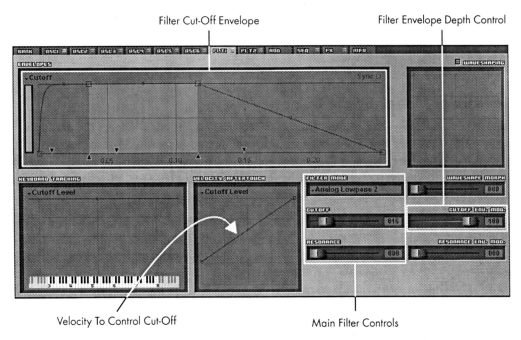

Velocity To Control Cut-Off   Main Filter Controls

Figure 6.3   Setting up Rhino so that the filter cut-off can be modulated by the
filter envelope and velocity.

sizers, the envelope depth control does not work to further open the filter from the selected cut-off frequency.

So for instance, if the filter is fully open and the envelope depth control is set to zero, there will be no effect on the filter. However, if the envelope depth is increased to 100%, then the envelope will reduce the cut-off frequency and only when the envelope is at its maximum will there be no effect on the filter. Equally, if the filter cut-off is at 50%, then this limit will not be exceeded when the envelope is at its maximum.

■ The third factor is the envelope itself—as the envelope depth control has been set to 100%, this means that the envelope level controls the filter cut-off directly (subject to the maximum level set by the cut-off control).

■ The final factor controlling the filter is velocity.

As with the attack time, the way that velocity controls the filter cut-off is set in the Velocity/Aftertouch window:

■ The top of the grid represents the cut-off frequency set by the cut-off slider.

■ The bottom of the grid represents a fully closed filter. Like the cut-off envelope, this modulation controller works to close the filter from the cut-off frequency selected by the cut-off slider.

In this patch, the cut-off level will be:

93

- unaffected at higher volumes, and

- cut a lower volumes—with a velocity of zero the filter will be half closed.

This setting means that the patch will sound brighter at higher velocities.

Summary of Modulation in Rhino

You can see that Rhino offers a huge number of modulation options and gives the musician detailed control over every aspect of a sound. The downside to this approach is, as always, complexity. If you want to change the way that velocity affects a patch you may have to change it in eight different places (six oscillators and two filters). If you are modulating a patch with several sources, this could become quite tiresome.

The other disadvantage to this option is a certain lack of flexibility. For instance you cannot set an oscillator acting as an LFO to modulate the filter's cut-off. Instead you would have to draw an LFO type envelope (or use one of the presets) to modulate the cut-off and achieve a wah-wah type sound.

There is also a different approach to assigning controllers under Rhino where this must be done by using MIDI learn to control the parameters.

We will look at some more practical modulation options in Rhino, specifically when creating FM sounds, in Chapter 7: Frequency Modulation Synthesis.

Modulation in Wusikstation and Cameleon

Wusikstation and Cameleon 5000 both adopt a "modulation matrix". This is a flexible grid where you can choose:

- the modulation source

- the modulation destination

- the minimum amount of modulation, and

- the maximum amount of the modulation.

Let's look at the Wusikstation matrix, which is shown in Figure 6.4, in greater detail. There are 36 slots for possible modulations—each slot has five controls:

- **Source.** This selects the modulation source, so for instance if you want to add vibrato this would be an LFO.

- **Destination.** This selects the target to be modulated—using our vibrato example, in Wusikstation a layer's pitch would be the destination (of an LFO modulation source).

- **Min and Max.** These select the minimum and maximum extent that the modulation source affects its destination—if the minimum level is set to be less than the maximum, then the modulation will negatively affect the destination (for instance, if velocity modulates a filter to make a sound brighter at higher velocities, the sound would get duller at higher velocities if the minimum is higher than the maximum).

■ **Amount.** This governs the extent to which the modulation source affects the modulation destination and also how it affects the destination. To elaborate, if an envelope is set to modulate the filter and the amount is set to 127, then the envelope may fully open the filter. If the Amount is set to 65, then the envelope will only partially open the filter. However, if the Amount is set to -127, then the envelope will operate to close the filter. The Amount control operates in conjunction with the Min and Max controls and so you may have to balance the three controls to get the effect you want.

This form a matrix gives considerable flexibility and control to the modulation routings.

Figure 6.4
Wusikstation modulation matrix.

MOD MATRIX					
SOURCE	DESTINATION	MIN	MAX	AMT	1
Mod Env 2	O1 Filter1 Freq	0	127	59	
Mod Env 2	O2 Filter1 Freq	0	127	79	
CC 1 ModWheel	All O and W Fine	0	127	58	
Mod LFO 1	All O and W Fine	0	127	92	
---	---	0	127	0	
---	---	0	127	0	

Wusikstation also gives another option which isn't immediately obvious from looking at the matrix—if two modulation sources are both set to control the same destination, then the effect of the sources is multiplied. So for instance, if an LFO and the modulation wheel are both set to control the pitch of an oscillator, when the modulation wheel is set to zero the LFO will have no effect (that is its effect will be multiplied by zero). However when the modulation wheel is set to its maximum extent then the LFO will have its full effect. This process allows greater control of the modulation. Wusikstation has other options, but this multiplication is the default.

Wusikstation also has three dedicated controls on each of the oscillator envelopes and modulation envelopes. These hardwire three destinations to the velocity control (as an example, one links the velocity to that layer's/envelope's level).

Modulation in Z3TA+

Like Wusikstation and Cameleon, Z3TA+ adopts the modulation matrix approach. However, as Figure 6.5 shows, the matrix here is more flexible/complicated and therefore gives you many more options as to how you want to apply your modulation.

Z3TA+ gives us five columns in the modulation matrix. Source, Range, and Destination act in a similar manner to the Wusikstation and Cameleon matrixes. There are two main differences with the Z3TA+ matrix: Curve and Control.

Figure 6.5 Z3TA+ modulation matrix.

Curve

Curve affects both how the source affects the destination and also the range of control. Some explanation might help.

Let's start by looking at the pitch curves—there are four choices: 1 semitone, 1 tone, 1 octave and 4 octaves. If you apply an LFO as a modulation source, then the curve will determine the maximum amount of pitch variation—so for a gentle vibrato you may select "1 semitone". This would give a vibrato range of +/- one semitone and so you might want to move the range slightly off the maximum.

The pitch curves do not have to be used exclusively with LFOs. For instance, you could equally apply a pitch curve in conjunction with the pitch envelope. Also, there is no rule that the modulation destination has to be pitch—you can use this curve in connection with any destination. And for completeness, you can use the other curves with pitch destinations—the pitch curves just make it easier to understand what the modulation is doing.

The next group of curves is the speed curves—these allow you to select curves which have greater effect either sooner in the curve (fast curves) or later in the curve (slow curves). Both of these curves have two options: positive and negative—if you're using the modulation wheel to control a filter's cut-off, then positive curves will open the filter and negative curves will close the filter.

The last group of curves we will consider are the linear curves—again these come in positive and negative flavors, but now we have another option: bipolar and unipolar. Bipolar means that the curve can generate negative *and* positive values. Unipolar means that the curve can be *either* negative *or* positive. Bipolar positive also has the effect of converting a bipolar source to unipolar.

Control

The second new feature of the Z3TA+ modulation matrix is the Control column.

Control allows a musician to "control" how the modulation source modulates the destination. So taking the example of an LFO acting as a vibrato source, if the modulation wheel is set as the control, then the depth of the LFO vibrato will be directly controlled by the modulation wheel (within the parameter set by the range control).

The range of controls is wide, ranging from velocity, to MIDI Control Code (MIDI CC) data through to key position.

96

Similar functionality can be applied in the Wusikstation matrix, it is just a bit more fiddly and takes two lines (which isn't a problem being as the Wusikstation matrix has 36 lines).

Z3TA+ Modulation Matrix Working in Practice

Here's an example of the Z3TA+ modulation matrix being used in practice. A similar (but not quite identical) functionality could be achieved in Wusikstation. Less functionality would be available in Rhino, Cameleon 5000 and Vanguard.

We're going to create a bright bass sound (which will take up quite a bit of space in the mix). If you've got the presets, then load up Bass Modulation.

This patch is based on two slightly detuned multi-mode (synchronized) oscillators (using the Vintage Saw 1 wave) running into a 36 dB low-pass filter. The filter is fully closed (requiring quite a bit of modulation to open it) and the resonance is set at quite a high level (10 dB). Finally, some reverb is added—the parameters are explained later.

If you manually input these parameters into Z3TA+ you will find that the patch doesn't make any noise. The silence is because the filter is fully closed. So the first thing we're going to do is to modulate the filter.

Filter Cut-Off Modulation

To open up the filter, I have set filter one as a modulation destination. If you set the Source to On, then the Range control will directly open the filter.

However, I wanted some velocity control here, so I selected Velocity in the Control column and I also chose the U-lin+ curve. Next I set the limits of the filter modulation, in other words the maximum and the minimum amount that the filter will be opened (so that the sound is neither too bright nor too dull). For this patch a Minimum of 26% and a Maximum of 67% felt right to me (left-click and drag in the Range to control the maximum and right-click and drag to control the minimum).

Finally, I added a bit of envelope control here—this adds a bit of bite to the sound by ensuring the filter opens a bit more at the start of the note and then becomes slightly duller. This behavior mimics real instruments.

To add the envelope control, I set EG1 in the Source column. Envelope one is used as a basic ADSR envelope (so Slope Time was set to the minimum and Slope Level to the maximum). Attack is as fast as possible (zero), Decay is set to 0.18ms and the Sustain Level is set to 70%. This gives a good sound to the filter modulation.

If you compare the with/without EG1 sounds, then without the envelope, the sustain portion of the note is considerably brighter. With the envelope, the start of the note has a more distinct attack and is less bright during the sustain phase—we will fix this dulling of the sound in a moment.

If you play this patch in the bass region it sounds fine. If you play the patch in the higher regions of the keyboard it sounds somewhat indistinct. Although this patch is a bass patch, I want to give it a bit more flexibility, for instance, so that it could be used as a stab (even though I am a keen advocate of designing patches for a specific purpose). This is where we start using key tracking.

As my modulation source I selected U-Note# with U-Lin+ as the curve and again Filter One Cut-Off as the destination. The range was adjusted to taste and in this instance a value of 38% felt right to me. This key tracking also has the effect of reversing some of the dulling effect of bringing EG1 as a modulation source.

Also note that if you want to use this patch as a stab you will need to increase the polyphony—I kept it at one to ensure the patch is not played in a way that would lead it to being muddy.

The filter is now highly dependent on the velocity of the incoming note and the high resonance value exaggerates the effect of the filter modulation.

Controlling the Volume

The amplitude envelope has no direct effect on the volume over time of this patch—it has the characteristic of an organ's envelope (so the sound is either on or off). Instead the volume of the patch is controlled by modulating the level of filter one.

To set up this modulation:

- Envelope Generator Two was chosen as a modulation source (and the envelope is set much like Envelope One, although the Sustain Level is slightly lower)

- the full Range was selected

- Velocity was engaged in the Control column, and

- the modulation destination is Filter 1 Level.

The volume modulation is now as sensitive as the filter modulation—this gives the patch a very playable feel to it.

Adding Some Drift/Detuning

So far the sound of the patch is a bit lifeless to my mind, so I added a bit of detuning. The oscillators were already detuned with reference to each other, so I added a bit of subtle vibrato. To make the effect more interesting, I set a different LFO speed for each of the oscillators—this ensures that the LFO does not set up a repetitive pattern that could become fatiguing on the ears.

To set this up:

- LFO one was assigned as the Source modulating oscillator one's pitch

- LFO two was set as the Source modulating oscillator two's pitch

- the one semitone pitch Curves were selected for each of the two assignments

- a sine wave was chosen as the wave for both LFO one and LFO two

- the Speed of LFO one was set to 0.58 Hz and the Speed of LFO two to 0.43 Hz

- the LFOs were applied to taste by tweaking the Range control in the modulation matrix—to my ears 9.6% sounded right for LFO/oscillator one and 13.7% sounded right for LFO/oscillator two.

The sound of the patch has now taken on a richness.

Adding Vibrato

This patch is already very playable, but I also wanted to add some controlled vibrato. To do this I:

- called up LFO three and set the wave to triangle and the Speed to 4.93 Hz

- assigned LFO three as the modulation Source which modulates the pitch of all oscillators

- set the Range to full and the Curve to Pitch > Whole Note, and

- set the modulation wheel to be the Control.

This allows the player to add vibrato without affecting the modulation introduced by LFO one and LFO two.

Reverb Without the Mush

As a final step I wanted to add some reverb, however, I was concerned that this could add some low end "mush", so I needed to ensure this reverb is controlled.

For this patch I used the Large Hall algorithm with the Size set at 75%, the Damping at 85%, Low at -3.6 dB, High at -1.97 dB and Wet/Dry at 75%. This gives a very noticeable wash of reverb, but if I cut the time, cut the mix level or added more damping I didn't like the effect, so did two things to control the reverb.

First I made the amount of reverb touch sensitive, in other words, at lower volumes the reverb will be proportionately less prominent. This is quite simple to achieve—in the matrix:

- the modulation Source is On and the Range is full

- the Control is Velocity, and

- the Destination is Reverb Level.

Now, with louder playing the reverb effect increases. However, there is a leveraging effect—more signal is being sent to the reverb because the signal is louder and secondly, the reverb itself becomes louder with harder playing.

The second thing I did was to gate the reverb (that is cut the reverb output signal) while the patch is being played. This is quite simple:

- first, a new envelope (in this case Envelope Generator Three) was engaged as the modulation Source

- envelope three was set to act and an ADSR type envelope with the Sustain level set to the full level—the only tweak was to set the Attack Time and the Release Time to 0.09 seconds

99

- the modulation Destination is Reverb Level, and

- the Curve is U-Lin-. The negative curve means that when a key is held, envelope three has a negative effect on the reverb level (in other words, it cuts the level completely).

While the notes are sounding, reverb will not be heard. As the notes are released, a halo of reverb is left. In this way the reverb adds effect, but doesn't make the mix muddy. The attack and release times could be less, but they are set just off zero to ensure that the transitions are not too unnatural to the ear.

Modulation in Surge

Surge takes a different approach to modulation that doesn't use a modulation matrix but still provides a system that is both flexible and intuitive.

For each of the modulation sources, there is a dedicated button. When you click on the button, you select the source so you can adjust that source's parameters—when you click on the button again, Surge goes into modulation assign mode. With this option, the sliders in all of the possible modulation destinations change their appearance: their slider caps turn blue and an immovable ghostly gray slider is added (although you won't be able to see the gray slider immediately).

Once in modulation assign mode, you can set the modulation by moving the (blue) slider that you want to modulate to its maximum modulation position. As you move the blue slider, it will reveal the gray ghost slider at the unmodulated starting point (see Figure 6.6).

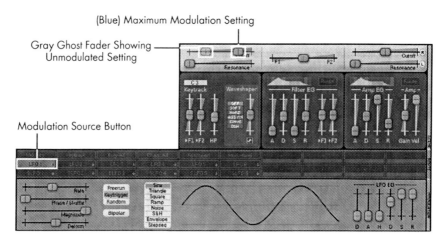

Figure 6.6   Setting up modulation in Surge. In this case, LFO 1 is modulating the filter cut-off—the ghost fader (showing the unmodulated position) and the maximum modulation fader are both highlighted.

Once you have set the maximum modulation to taste, you can click on the modulation source button again and the blue sliders will disappear. For the function that is being modulated, a blue background will show behind the slider to indicate a modulation has been set.

The modulation that has been set is relative to the level of the slider in non-modulation assign mode, so if you adjust a slider that is being modulated, the modulation will be affected since its amount is set relative to the base setting of the control. As long as you've got your head around this concept, Surge's approach is a very intuitive way to set up your modulation.

You will also see that Surge has two dedicated envelopes for each scene: a filter envelope and a volume envelope. This makes the use of envelopes much faster for these two functions.

Chapter 7

Frequency Modulation Synthesis

This is not a science book, and nor is it a history book, so I will not be explaining the principles behind how FM synthesis works. Instead, I will talk in this chapter about the practical aspects of creating sounds using frequency modulation principles. These principles also apply when creating sounds with phase modulation which is similar to FM. Of the featured synthesizers, only Z3TA+ offers phase modulation.

If you want to understand more about FM theory and its history (and I would encourage you to dig deeper) then I suggest you start by using the following terms to search the internet:

- **John Chowning.** Dr Chowning was the guy who first got to grips with the concept of FM synthesis: all musicians owe a great debt to him.

- **DX7.** The Yamaha DX7 was the first mass market FM synthesizer (if you have heard any pop music produced during the 1980s then you will have heard a DX7).

- **FM synthesis.** This term should be self explanatory.

Elements to the FM sound

For many people, FM synthesis falls into the category of "too difficult"—it seems both cumbersome and complicated giving unpredictable results.

However, FM is comparatively simple in concept—it just has a lot of possibilities which make it appear quite complicated. It also requires a different mindset from that needed for subtractive synthesis. So forget every way of working we have discussed so far and let me explain.

An Example Patch Structure

Before going any further, I want to have a quick look at an example patch structure. This should give a context to how an FM patch may be constructed.

It is quite usual for an FM patch to include several elements. For instance, a classic electric piano type patch will have two main elements—the bell tone (heard when the key is struck) and the sustain sound. From this foundation the sound can then be developed (as Figure 7.1 shows).

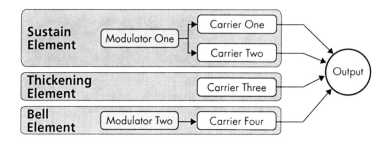

Figure 7.1   An example of the structure of an FM piano.

In the figure you can see:

- A first modulator/carrier group (these terms are explained below) creates the sustain sound.

- A second carrier (modulated by the first modulator) doubles the sustain sound created by the first carrier—using the same modulator for both carriers is more efficient but restricts the flexibility of the sound.

- A third carrier (this time without a modulator) is used to thicken up the sound—the three carriers together create a cohesive foundation for the sound.

- Another modulator/carrier pairing then creates the bell sound.

One of the difficulties when creating a sound of this nature is to ensure that the two separate parts of the sound (in this case the sustain part and the bell part) work together to create a convincing sound.

Operators

Instead of having oscillators, FM synthesizers have "operators". The first mass market FM synthesizer, the Yamaha DX7, had six operators—each of which was a sine wave. To create an FM sound, two operators are needed—a modulator and a carrier.

A modulator can be thought of as acting in a similar manner to a low frequency oscillator (LFO). It modulates the frequency of the carrier—the carrier is the operator that is connected to the output and is therefore heard. However, the frequency of the modulator is much higher than would be conventional for an LFO. Indeed, the frequency of the modulator is in the audio spectrum so the modulator could be heard if it were to be connected to an output.

The effect of modulating a carrier with a signal in the audio spectrum is that the effect is not heard as vibrato, but instead the tone of the carrier is changed, quite often significantly.

Software synthesizers which have an FM function (such as Rhino, Surge, Wusikstation, and Z3TA+) allow the option of including different waves as operators. These different waves give different sounds. You could probably spend a lifetime without exhausting all of the possibilities offered for FM synthesis using sine waves alone. To my mind other waves just give too many complications without giving such pleasing results, so are not a practical proposition—your opinion on this matter may be different.

The DX7 had six operators—these were arranged in algorithms. Rhino and Wusikstation also have six operators, however, their routing options are fully flexible so there is not the restriction imposed by hardwired algorithms. Z3TA+ also has six operators, but its routing options are slightly less flexible. Surge has many FM options and can allow up to 24 operators to be used.

Filters

The original FM synthesizers did not have filters. All of the synthesizers featured in this book which offer some form of FM also give the opportunity to filter FM sounds. This gives yet more possibilities. As you will see later, it is also possible to create hybrid FM/subtractive patches to take advantage of the best of both worlds.

Subtractive synthesis can be characterized as creating "fat" or "warm" sounds, where digital is often characterized as being "thin" and "cold". Both characterizations are generalizations which have elements of truth and falsehood (it really all comes down to the quality of the programming).

However, in some circumstances:

■ a subtractive sound can dominate a broader element of the frequency range than an analog sound, but

■ an FM sound can be brighter/sharper, cutting through a mix.

Provided you have programmed your sounds in an appropriate way, this can means that FM and subtractive sounds can be the perfect compliment for each other, with the characteristics allowing the respective sounds to work together perfectly into a mix.

Controls

The essence of the sound produced by an FM synthesizer is not simply a combination of the individual elements, but the interaction between the modulator(s) and the carrier(s) over time. As one element changes in relation to another there are tonal shifts which gives FM its unique sound. The aspects that affect a sound are:

■ the frequency of the modulator relative to the carrier—as a general rule, the higher the frequency of the modulator relative to the carrier, the brighter (or more metallic) the sound, and

■ the amount of modulation: this is controlled by the output level of the modulator—the greater the output, the greater the effect of the modulator (again, this affects the brightness).

To achieve the constant shifts and design the desired sound, each element must be controlled: usually (but not always) the pitch relationship between each operator is fixed (and controlled by the keyboard), but the level of each operator is dynamic. There are several tools in our armory to bring about these controls—envelopes, velocity scaling and key scaling—we have used them all in subtractive synthesis.

Envelopes

Much of the character of a sound—any sound—is captured in the note's attack. Conventional ADSR envelopes may not give sufficient range of control and so since the earliest FM synthesizers, more complicated envelopes have been used. Rhino, offers highly flexible envelopes (which offer a much greater degree of control than was available on the original DX7).

FM sounds where the modulators are static can often be quite harsh and fairly uninteresting.

When sounds with many layers are created, the envelopes are key to achieving the necessary control over the sound. Take the example of the electric piano (set out above): when constructing the envelopes, the sustain portions should (as would be expected) sustain. The bell elements of the sound should have quite a percussive attack and a swift decay. However, the decay should be sufficiently smooth that the change between hearing the bell element and the sustain elements is imperceptible to the listener (and the musician).

As you will see later, the attack signature of a sound can be readily controlled by having different attack times for the modulator and carrier envelopes. For instance, to create the "spit" of a brass note, a slower attack time can be selected for the modulator (when compared to the attack time for the carrier).

Velocity Scaling

When creating "pure" FM sounds (in other words, when not using filters) the tone is often controlled by using velocity scaling to control the output level of the modulators. The output of the whole patch is controlled by using velocity scaling to control the level of the carriers.

Even more than is the case for subtractive synthesis, velocity scaling does not have to have a range from zero to the maximum. For instance if velocity scaling is controlling the tone of a sound, then you may want to set one end of the velocity range so that the sound is dull and the other so that the sound is bright—dull and bright may not correspond with no modulation and maximum modulation.

Velocity scaling can give the musician immense amounts of control over an instrument, perhaps to a level where the synthesizer can come close to mimicking some of the behaviors of a real instrument.

Key Scaling

With subtractive synthesis, key scaling is usually only used to open up the filter a bit when higher notes are played. With FM synthesis, key scaling is far more important.

It is very easy to create metallic sounds with FM synthesis. It is harder to create usable metallic sounds. One of the keys to designing FM sounds is to ensure that the overtones are appropriate. It is quite easy to create a sound which works within a limited key range, but does not work outside of that range. Often this will manifest itself as a sound that begins to sound too harsh. The harshness can be addressed in two ways:

■ the modulation can be reduced—this may make the sound work outside of the first range, however, it will probably affect the original sound, perhaps making it too dull

■ the level of the operator can be reduced in certain key ranges—this is more likely to obtain the desired result. This is what we mean when we talk about key scaling (or key level scaling as it is sometimes called).

Tuning

With FM sounds tuning can be quite a challenge. First, the pitch can be affected by the interaction of the modulator and the carrier. Secondly, the resulting waveforms can include many harmonious and inharmonious elements—if the inharmonious elements are predominant, then the resulting sound (especially if a chord is struck) may not be pleasant.

Getting to Grips with FM Programming

The essential element of the FM sound is a modulator and a carrier working together. Sounds can have more than one modulator: there may be several carriers each with their own modulator or several modulators modulating one carrier—these possibilities are described below.

Combining Operators

As we have mentioned, there are many ways that operators can work together to create a sound. For the sake of clarity, let me explain the terms I am going to use in this chapter.

Simple FM

Simple FM is where one modulator drives one carrier (see Figure 7.2).

Figure 7.2 A simple FM stack where one modulator drives one carrier.

This arrangement is the very essence of FM sound. Most of the classic FM sounds can be built around this arrangement, although often several simple FM stacks will be layered together to achieve a thicker sound. Alternatively, some patches may be built around several simple FM stacks providing different elements of the sound: the classic example of this is the FM electric piano sound where the bell and the sustain portions can be built from separate simple FM stacks which are then layered.

Often one modulator/carrier stack alone will give a weak sound, much as a single oscillator in a subtractive synthesizer can give a weak sound. Modulators and carriers can be doubled to thicken up a sound—again think of how oscillators are doubled in a subtractive synthesizer.

The examples in this chapter use Rhino, however, three of the other synthesizers (Surge, Wusik-station, and Z3TA+) have FM capabilities. All of these other synthesizers can easily implement simple FM stacks and can have up to three simple FM stacks arranged in parallel. However, Z3TA+ does not have the routing options to allow either of the parallel operator arrangements described below.

Surge takes a different approach to FM and offers some dedicated FM algorithms. These can make setting up FM sounds much easier, and it also gives the opportunity for more routing options. Accordingly, I have dedicated Chapter 7 Supplement, which follows this chapter, to creating FM sounds in Surge.

Parallel Carriers

With parallel carriers, one modulator drives two or more carriers (see Figure 7.3).

Figure 7.3 A parallel carrier arrangement with one modulator drives two carriers.

Parallel carriers allow the effect of two simple FM stacks to be achieved by using three operators rather than four. This leads to greater programming efficiency and marginally less CPU load.

The downside of parallel carriers becomes apparent if you detune one carrier (say by a couple of cents) to get a thicker sound. While the sound is thicker, there is also phase cancellation which is primarily heard as a rise and fall in the volume.

However, if you use two simple FM stacks and detune both the modulator and the carrier in one stack, a much richer sound can be achieved without such apparent phase cancellation (but you are using more operators and so losing one advantage of parallel carriers).

Parallel carriers do not need to be at the same pitch. For instance, if you take one carrier at the base frequency and another seven semitones higher, you have the basis for a Wurlitzer type electric piano sound. Equally this arrangement can yield many wooden (as in tuned wooden percussion) type sounds.

As a side note, just because two unmodulated parallel carriers sound odd when they are played together, that does not mean they will not work together when they are both modulated by the same modulator.

Parallel Modulators

With parallel modulators, two or more modulators drive one carrier (see Figure 7.4).

Figure 7.4 A parallel modulator arrangement where two modulators drive one carrier.

Parallel modulators offer far more complicated and rich sounds than are available from a simple FM arrangement. However, more options give more complications: the relationship between the two modulators will have a considerable effect on the sound.

If the modulators are of the same pitch, then the effect of two modulators will be to enhance the amount of modulation. Tonally this means that the brightening effect of the modulation will be increased.

Very simplistically, if the modulators are all set at intervals which correspond to integer multiples of the carrier's frequency, then the sound with parallel carriers will give a harmonious tone. Different intervals can give many different tones—some of which may be useful in a musical context, other of which may fall more to be considered as sound effects.

Another simple, but effective use for parallel modulators is to apply different envelope attack times—used in conjunction with differently tuned modulators, this can give a very natural sound. Again, taking the example of a simple electric piano:

- the first modulator could have a very fast attack and a short decay—this could be tuned to a higher pitch than the carrier to give the bell sounds, and

- the second modulator could have a slower attack and a sustain portion—this could be tuned to a pitch below the carrier to give a more t/wooden type tone.

If two modulators are very slightly detuned, then this will lead to interference between the modulators (in the same way as would be the case if two oscillators in a subtractive synth are detuned). However the net effect is that the amount of modulation (and therefore the tone) will constantly shift as the operators cancel each other out. Depending on the particular patch, the results may or may not be desirable.

Cascade

With a cascade, one modulator drives another modulator which itself then drives the carrier (see Figure 7.5).

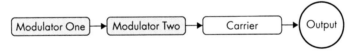

Figure 7.5 A cascade where one modulator drives another modulator, which itself then drives the carrier.

With a cascade, the first modulator outputs a sine wave. However, the second and subsequent modulators are themselves being modulated. Accordingly their output is a different waveform. Cascades can produce the richest and most complex FM timbres: a similar (but less controllable) effect can be created in a simple FM stack by modulating with waveforms other than the sine wave.

Although not illustrated in the graphic, it is possible to have more than three operators in a cascade, however the results then become less predictable and there is a greater tendency to producing noise.

Surge, Wusikstation, and Z3TA+ can all readily implement FM cascades—indeed, Z3TA+'s architecture was specifically designed to allow this sort of FM arrangement.

Building a First FM Patch

As a first step towards programming FM, let's look at an FM patch. Once we have done this we will then go back and look at the elements of the FM sound in greater detail. If you have the patches, load the bank Rhino Chapter 7 FM Synthesis Bank, and call up the patch FM Bass (which should be at the top of the list).

All of the example patches in this chapter are built with Rhino. You will make your life much easier if you familiarize yourself with the Rhino user manual. We are also going to be using many of the modulation concepts which were discussed in Chapter 5, so please check out that chapter before you get too deeply into this chapter. As I have already mentioned, we will look at patches created in Surge in the supplement to this chapter.

If you play FM Bass you will hear only one operator: operator one, which is a sine wave. This patch will end up as a three operator patch (and two further operators can be called into action for illustrative purposes).

Setting Up the Patch

For this patch, each of the five operators has a sine wave selected. An envelope has been drawn (it is the same envelope for all operators) with an immediate attack and an exponential decay, so the sound initially decays swiftly but the rate of decay slows. This envelope mimics the envelope of a piano or guitar. For this example there is no key scaling or velocity scaling. This is intended as a bass sound, so please play the patch in the lower ranges (although the patch is quite playable at higher pitches).

If you look at the Rhino matrix, you will see that some of the numbers are "grayed" out. This means that the operator is working, but that routing has temporarily been muted. To turn the routing on again, simply right-click on the grayed out number. If you didn't know about this, go back and read the Rhino user manual.

For anyone who has not purchased the patches but who wants to create this patch you will need the following settings (if you have the patches, skip this section):

- Level envelope (for all oscillators): fastest attack time, decay time 2 seconds (after which volume = zero), curvature control set to approximately 25 (see Figure 7.6).

- All of the oscillators have sine waves.

- Oscillator one has its coarse pitch set to zero, its Raw output should be set to 100, it is modulated by oscillator two and oscillator three, both modulation Levels should be set to 60.

- Oscillator two has its Coarse Pitch set to -19, it does not have any output (so it only acts as a modulator), it is modulated by oscillator three and oscillator four, both modulation Levels should be set to 100.

- Oscillator three has its Coarse Pitch set to -12, its Raw output should be set to 100, it is modulated by oscillator four—this modulation level should be set to 100.

■ Oscillators four and five both have their Coarse Pitch set to -12 and their Raw outputs set to zero, oscillator four is modulated by oscillator five—this modulation Level should be set to 100.

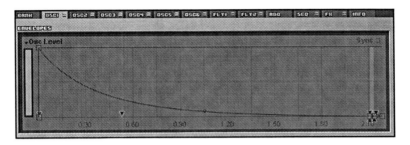

Figure 7.6 FM Bass envelope.

Once you have set this patch, all of the modulations should be switched off (by right-clicking in the matrix). Only the raw, unmodulated, output from oscillator one should remain.

Initial Modulation: Simple FM

Once you have loaded FM Bass and made the mutings so that only operator one is heard, listen to the sound of that operator. Play some notes in the bass region: you will hear a pure sine wave. Then right click where the figure "1" shows in Figure 7.7 (that is where the Osc 1 column and the Osc 3 row intersect)—this will engage operator three as a modulator for operator one, and the figure will no longer be grayed out. You will hear that the sound immediately becomes:

■ louder

■ lower in pitch

■ darker in tone.

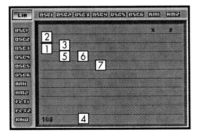

Figure 7.7 The routing matrix for FM bass.

Operator three has its pitch set one octave (12 semitones) below operator one's pitch. When an operator with a lower pitch modulates a higher pitched carrier, the sonic result is a louder, lower, darker note.

Switch off that modulation and switch in the modulation indication by the number "2" in the image (so right-click where the oscillator one column and the oscillator two row intersect). This will engage operator two as a modulator for operator one. In comparison to the sound when operator

111

three was the modulation source, this sound will be even lower in pitch (and may subjectively be darker in tone).

The modulator in this case (operator two) has been pitched 19 semitones (an octave and a fifth) below the pitch of the carrier. You should note that as a side effect, the perceived pitch of the sound when the modulator is pitched below the carrier is more likely to be governed by the modulator rather than the carrier.

Go wild—bring in operator three as well by right-clicking as indicated by the figure "1". You will hear that the tone changes quite significantly and the pitch again feels lower.

Developing the Patch: Cascading Operators

Switch out operator three as a modulation source for operator one: you should now have a simple FM sound of operator one being modulated by operator two. Let's cascade the modulators—set operator three to modulate operator two (which is modulating operator one). Right-click (as indicated by figure "3") where the oscillator 2 column and the oscillator 3 row intersect.

You will hear quite a radical change in tone. The sound becomes much brighter and more aggressive without losing its bass weight.

Now also add operator 3 as a modulator to operator one (right-click where indicated by the number "1").

Adding Weight

To really add some weight to this sound, operator three can also act as a carrier. Right-click where indicated by number "4" and you will hear the sine wave generated by operator three being added directly to the output, as well as acting as a modulator to operators one and two.

Dominant Operators

Take some time to listen to the interaction of these three operators. The dominant element of the patch is the three cascaded operators—operator three modulating operator two which in turn is modulating operator one. This is also the dominant element in the pitch. If you play this patch with other patches it will sound in the wrong key: this is the effect of operator two. To reach regular pitch, all of the operators should have their pitch raised by a fifth (in other words seven semitones).

First Patch: Next Steps

You could take this patch further and add more modulators. While interesting, I preferred the simplicity of the three operator patch. However, you may disagree. In any case, here are a few simple additional modulations you could add:

- Operator four could be set to modulate operator two (number "5" in the graphic). This adds an interesting element to the tone.

- Alternatively, operator four could be added to the start of the cascade and so modulate operator three (right-click where indicated by the number "6" in the graphic). On its own I don't particularly like this tone, but I'm sure I could find a use for it.

- Finally, operator five could be added at the start of the cascade (right-click where indicated by the number "7"). I don't like this sound—however, it does illustrate the downside of adding the additional operators: the noise. The patch goes from having a rich sound to having annoying elements of noise.

Velocity and Key Scaling

If you wanted to take this patch further, then by applying some velocity scaling to operator one, the volume could be controlled by velocity. The tone can be controlled by tweaking the velocity response of some of the modulators—ideally operator three.

However, the challenge here with velocity scaling comes because the effect of the modulators is to make the sound darker—reduce the volume of the modulators and the sound gets less dark (a good thing in this context). Remove the modulators and you are left with a sine wave at a higher pitch than the fundamental of the bass note (a bad thing). Also, operator three is both a carrier and a modulator: if you cut its volume too radically you will lose even more of the weight of the sound. As with all sounds, your job is to balance the various elements to produce a musically useful sound.

Building Blocks of FM Sound

I now want to walk you through the basic elements of the FM sound. For this you will find it much easier if you have the accompanying patches. The bank accompanying this chapter, Rhino Chapter 7 FM Synthesis Bank, contains around 60 Rhino patches. If you don't want to buy the patches you can still follow the chapter. However, you may find it harder to pick up some of the nuances in the detailed patches towards the end of the chapter.

1:1 Ratio—FM Using Different Modulators

This first group of patches illustrate the basic tones that are available in a simple FM stack by modulating a sine wave with various different waveforms. I have used five different waves—Rhino has over 100 waves so there is a lot of room for experimentation if you are curious.

For these examples the modulator in each patch is touch sensitive. This will give you an idea of the effect of the level of the modulator on the tone of the patch. You should also note that for these patches the envelopes are very basic (the operator levels do not change over time).

For those of you without the patches, you can set up Rhino for this group of sounds in the following way:

- operator one (to act as the carrier)—Coarse Pitch set to 0

- operator two (to act as the modulator)—Coarse Pitch set to 0

- both Level envelopes—attack time: 0, no decay phase, sustain at full level

- amount by which modulator modulates carrier = 100

- the modulator is fully velocity sensitive. To set this, go to operator two's page—in the center you will see a box labeled Velocity/Aftertouch. Choose Level from the drop-down (if this is not already selected) and then drag the left hand square on the velocity scale to the bottom left corner. Ensure that the right square is in the top right corner.

The waveforms used are suggested by the patch names.

Sine to Sine

This is the most elementary FM configuration. The sound produced has a reedy/woodwind quality that is perhaps reminiscent of an organ's tone.

Triangle to Sine

When a triangle wave is set as the modulator the result is a much brighter tone than when the sine wave is used. However, at lower levels of modulation, either when less velocity is used or if the output of the modulator is reduced, the sound of Sine to Sine and Triangle to Sine are similar.

Square to Sine

Square to Sine gives a brighter sound again, and the overtones generated by frequency modulation can be more clearly heard. However, this patch shows an interesting behavior—at higher levels of modulation (in other words at higher velocities for this patch) the sound gets thinner and quieter.

Saw to Sine

Like Square to Sine, Saw to Sine sounds thinner and quieter with higher levels of modulation. With higher levels of modulation the sound also tends towards noise: generally this may be unwanted, however, it may be useful, for instance when creating a woodwind type patch.

At mid levels of modulation, the sound is quite warm and rich and is reminiscent of an overdriven-type sound.

With the exception of Sine to Sine, these patches all exhibit (to a greater or lesser extent) the following behaviors:

- at low levels of modulation, the sine wave alone is heard

- as the effect of the modulator starts to become apparent, two separate elements are heard—the sine wave and some higher frequency "distortion"

- at mid levels of modulation two elements are present but the sound becomes warmer

- at higher levels of modulation only one element of sound is heard (largely because the distortion sound drowns the sine wave)

- as the levels of modulation increase further, the sound tends to noise and becomes quieter.

Brown to Sine

Brown to Sine takes a different approach and uses noise (in this case brown noise) as the modulation source.

This arrangement gives a bit of a strange result. The sine wave is clearly audible and does not seem to be affected much by the process. However, the brown noise is amplified by using it as a modulation source. If you switch off the brown noise as a modulator and engage it as a carrier you will hear the natural tone of the noise (being less strident than when used as a modulator) and the volume will be reduced.

1:1 Ratio—FM Using Different Carriers

The next group of patches illustrates the basic tones that are available in a simple FM stack when different waves are modulated with a sine wave.

For these examples the modulator in each of the patches is again touch sensitive. This will give you an idea of the effect of the level of the modulator on the tone of the patch. As before, the envelopes are very basic (the operator levels do not change over time).

Sine to Triangle

When the sine wave modulates the square wave, the sound gets brighter. It takes on the quality of a slightly nasal sawtooth wave.

Sine to Square

When a sine wave modulates the triangle wave, the resulting sound is closer to two pulse waves which have been phase synchronized—the tone, as you would expect, is brighter and harder but still very musical.

Sine to Saw

The sawtooth wave is already harmonically rich. When modulated by a sine wave, the effect is quite subtle: at the maximum amount of modulation, the sawtooth has a thinner, perhaps slightly sharper, quality.

Sine to Brown

The effect of the sine wave modulating brown noise is to make the noise output brighter and louder. If you want a brighter noise source, it may be quicker to choose a different color noise (for instance, pink noise).

Comparison: Sine Wave as Carrier or Modulator

While rather a sweeping generalization, it can be heard that a key difference between the patches where the sine wave is the modulator and the patches where the sine wave is the carrier being modulated by a different waveform is the consistency of the output signal. Where the sine wave is the modulator, one sound is heard (irrespective of the level of the modulator). Where the sine wave is the carrier, there are generally two sounds, the sine wave and some distortion—there is also a tendency to noise.

However, do not let these comments rule out the other wave forms as modulators (or carriers for that matter). These different waveforms give different shades of tone—their results will only be unpredictable until you become familiar with their operation.

1:1 Ratio—FM Using Same Modulator and Carrier

This last group of patches illustrates the tones that are available in a simple FM stack when an operator is modulated by the same waveform. You should note that the results here differ from those that will be achieved by creating a feedback loop in an operator. With a feedback loop the sound will get brighter but, as might be expected, at higher levels of feedback there is far more of a tendency to noise.

For these examples the modulator in each of the patches is again touch sensitive. This will give you an idea of the effect of the level of the modulator on the tone of the patch. As before, the envelopes are very basic (the operator levels do not change over time).

Triangle to Triangle

As may be expected, when a triangle modulates a triangle, the tone becomes sharper and more "plucked". As the modulation level increases or decreases, the tone takes on the quality of a sound where the pulse-width of the oscillator is being modulated. This could provide a quite usable musical tone.

Square to Square

As the square wave modulates another square wave, the resulting sound becomes thinner and sharper. At higher ends of modulation, noise is introduced.

Saw to Saw

As the sawtooth wave modulates another sawtooth wave, the sound gets noisier and quieter. From a practical perspective, there is little to commend the use of this combination.

Brown to Brown

Your ears may be better than mine, and you may have a better listening environment, but to my mind there is no significant difference in tone when brown noise is modulated by brown noise.

Practical Uses for the Same Modulator and Carrier

As can be heard the range of musical options that is available when the same wave is used for both the modulator and the carrier in a simple FM stack is far more limited than may at first be expected. For this reason, the rest of this book will use sine waves as FM operators (except in one or two cases), however, this should not discourage you from developing a greater understanding of the sonic possibilities offered by different waves.

All of these examples have been illustrated using Rhino but there is no reason why you shouldn't create similar sounds with Z3TA+. Equally Z3TA+ offers different options from Rhino—for instance different factory waves and more wave shaping options—there is no reason why you should not use these changes for creating FM tones. You could, of course, use Surge as I will do in the supplement to this chapter.

1:1 Ratio—FM Using Parallel Operators

This group of patches illustrates the tones that are available when parallel operators are used. For these examples the modulator(s) in each patch is/are touch sensitive. This will give you an idea of the effect of the level of the modulator(s) on the tone of the patch. As always, the envelopes are very basic (the operator levels do not change over time).

For reference, you may want to listen to Sine to Sine before you listen to these patches.

Sine2 to Sine

Sine 2 to sine is built around one sine wave with two parallel modulators at the same pitch as the carrier.

Two (equal pitched) sine waves both modulating another sine wave at 50% will give the same sound as one sine wave modulating another at 100%. The advantages of having parallel modulators are:

- the carrier can be modulated to a greater extent than is possible with one modulator (thereby giving the possibilities for a brighter tone still)

- the modulators can be tuned differently, giving greater sonic options (this is discussed in greater detail below), and

- the different modulators can have different envelopes allowing them to control the carrier in different ways at different stages of the note—often different envelopes will be used when the modulators are tuned differently.

Sine to Sine2

Sine to Sine2 is a basic parallel carrier arrangement where one modulator modulates two carriers.

There is little tonal difference between Sine to Sine and Sine to Sine2—the key difference is the increase in volume. Parallel carriers are most useful where the carriers are tuned to different frequencies (for instance, an octave apart).

1:1 Ratio—FM Using Cascading Operators

Moving on, let's look at some cascading operator patches and also consider the role of operator levels in cascades. Again, the envelopes used here are very basic (the operator levels do not change over time).

Sine to Sine to Sine

This first patch is the most simple cascade—operator three modulates operator two which then modulates operator one which then feeds the output.

Operator two and operator three are both velocity sensitive. Therefore, as the velocity increases, operator two will modulate operator one (the carrier) to a greater extent. However, operator two will also have its wave shape changed by operator three so as the velocity increases, two factors are changing the tone.

At the maximum extent of modulation, the tone is brighter than Sine to Sine. This sound could perhaps be used for a clavinet or a bass type patch. In the mid range, there is a woodwind type tone, reminiscent of a clarinet or an oboe.

117

Sine to SineX to Sine

With this second cascade patch, the effect of operator two on operator one is constant and is not controlled by velocity (although operator three is still velocity sensitive). What changes is the wave shape of operator two which is controlled by the extent to which it is modulated by operator three.

At lower volumes, the sound is the same as that produced by Sine to Sine at maximum velocity (since in this case operator two is not being modulated). The effect of fixing the relationship of operator two to operator one is to give this patch a generally brighter tone which is controlled by velocity

SineX to Sine to Sine

With this third cascade patch, the effect of operator three on operator two is constant and is not controlled by velocity. However, operator two is controlled by velocity and so velocity changes the extent to which operator one is modulated by operator two.

In the mid range this patch has a more woody tone than Sine to SineX to Sine.

Sonic Nuances of the Cascade Patches

At maximum modulation (that is at maximum velocity) all three of these patches give the same tone.

However, at lower volumes, Sine to Sine to Sine gives the least bright sound (since both operator two and operator three will have their levels reduced. While Sine to Sine to Sine may appear to have most flexibility as two operators are controlled by velocity, it cannot reproduce some of the mid range tones of the other two patches and so the velocity control (which could equally be any other form of control such as keyboard scaling) works to the detriment of the sound.

The last two patches illustrate how fixing the relationship between operators at different parts of the cascade chain can produce markedly different sounds.

More Unusual FM

At this point I want to look at a few examples of the more unusual applications of FM. None of the following four patches uses any form of velocity or key scaling.

Mock Sawtooth

It is possible to create the sound of a sawtooth wave with a sine wave and some FM. You can do this by setting an operator with a sine wave and feeding the operator back into itself—you will see that I have set a value of 57. If you want to compare this constructed sawtooth with the real thing, in Mock Sawtooth oscillator two has been set with a sawtooth wave (and muted). You will hear that the constructed version is quite close.

I think it is unlikely that you will want to use FM to create a sawtooth wave. However, you may want to be aware of this possibility. More likely you may find it preferable to replace a sine wave creating a sawtooth with a sawtooth wave.

Sort of Square

In a similar manner to Mock Sawtooth, you can create a sort of square wave sound from sine waves. To achieve this, the modulator should be set an octave above the carrier and the modulation should be set to full. Finally I set the feedback in the modulator to 21. Again, to compare this constructed wave with the real thing, in Sort of Square I have set up oscillator three with a square wave and muted it. You will hear that the real square is a bit brighter than this constructed wave.

We have now looked at two recreations of standard synthesizer waves. However, please do remember that at the end of the day this is rather pointless exercise—whether or not we can create a square wave or a sawtooth wave is not significant. What is important is whether any sound will work in the patches we will build.

Many Mods

We have already mentioned parallel modulators, and will discuss this idea further when we look at patch building later in this chapter. Many Mods is an example of a carrier being modulated by five modulators all of the same frequency. In essence, this is the Sine2 to Sine patch taken to the extreme.

I think the most noteworthy thing about this patch is how un-noteworthy it is. I cannot see any practical situation where you are likely to try to construct a patch of this manner.

Many Carriers

Many Carriers takes Sine to Sine2 to the extreme: operator six modulates operators one to five which are all of broadly the same pitch. You will see that each operator has been slightly detuned. This has the effect of thickening the sound a bit, but it also has the side effect of creating phase cancellations which lead to the fluctuating volume you will hear.

If you have the patches you will see that some feedback in operator six has been muted out. If you right-click this to activate the feedback, you will hear a change in tone—you will also hear that the effect of the volume fluctuations becomes more pronounced (and more intrusive).

Again, this patch is more important as an example of something you probably don't want to bother pursuing as it produces an uninteresting sound.

Varying Carrier : Modulator Ratio

I have already touched on the effect of having the modulator and carrier at different frequencies. At this point, I feel I am already running out of ways to describe the different nuances which arise with different modulator/carrier relationships. For this I apologize, however, we are reaching a point where words alone cannot convey sounds. If you haven't done so already, I suggest you get hold of Rhino (or a demo of Rhino) and work through some of the examples given below. It will yield far better results than reading this book alone.

This section describes a range of sounds that can be achieved with a simple FM stack where the modulator is at a different pitch by reference to the carrier. There are also a few parallel operator examples. I should point out that the carrier/modulator pitch intervals illustrated here are not the only options, and there is no reason for you to stick with the semitone intervals that I have used.

On the subject of tuning, I have described the interval between the modulator and the carrier in terms of semitone steps. I have done this because it is the most immediate way to describe the interval. Many other people talk in terms of carrier : modulator ratio (or C:M ratio) where the modulator frequency is a multiple of the carrier's frequency, so for instance:

- with a 1:2 ratio the modulator is 12 semitones higher than the carrier (the modulator would be twice the frequency of the carrier)

- with a 1:3 ratio the modulator is 19 semitones higher than the carrier (an octave plus a fifth—the modulator would be three times the frequency of the carrier)

- with a 1:4 ratio the modulator is 24 semitones (or two octaves) higher than the carrier.

I think you get the idea. If you right click on the Pitch Coarse slider in Rhino it will display the pitch as a multiple of the base frequency (see Figure 7.8)—this is an easy way to set a ratio as you are working with frequency ratios rather than semitones. This is also a more convenient way to sweep through frequencies without being bound by the semitone steps of the coarse pitch slider.

Figure 7.8 Setting oscillator pitch as a multiple of the base frequency in Rhino.

You will notice that the integer ratios give a cleaner (less clangourous) form of FM. In particular you will find that 1:integer C:M ratios will not give you that classic metallic sound.

Let's look at what can be achieved in a bit more detail. For these simple FM patches, there is no velocity sensitivity.

Sine -48 to Sine

In Sine -48 to Sine the modulator is pitched considerably below the carrier—in this instance it is pitched four octaves below, or if you want to consider the relationship of the frequencies, the carrier is approximately 16 times the frequency of the modulator. In this case, the modulator is acting more as a fast LFO than as an FM modulator. As would be expected, the result sounds more like very fast vibrato rather than a usable musical sound.

Sine -36 to Sine

When the modulator is pitched three octaves below the carrier (which gives a C:M ratio of 8:1) a usable tone starts to develop. The tone is engine like in the lower ranges and almost has a vocal quality in the upper ranges. The pitch of the note in comparison to the carrier's pitch is low.

Sine -24 to Sine

At two octaves below the carrier (C:M ratio of 4:1) a richer tone emerges which has more weight than the tone produced by Sine -36 to Sine or Sine -12 to Sine (which is considered below).

Sine -19 to Sine

At nineteen semitones below the carrier the C:M ratio is 3:1. The tone of this combination is very similar to that given by Sine -24 to Sine.

Sine -12 to Sine

As mentioned above, Sine -12 to Sine (C:M ratio 2:1) has a thinner tone than Sine -24 to Sine, but it does have a thicker tone than Sine to Sine.

However, if you just compare the sounds (of Sine to Sine, Sine -12 to Sine, and Sine -24 to Sine) by playing the same note you may give yourself an unrealistic impression of the sonic differences. Play Sine to Sine, then play Sine -12 to Sine one octave higher, and then play Sine -24 to Sine another octave higher still. You will still hear differences but they will be less pronounced.

Sine to Sine

We have already looked at this patch (and have already considered many C:M 1:1 ratio patches)—a tweaked version (to remove the velocity scaling) is included here only to give a comparison with the other modulator/carrier relationships. When compared to the earlier tones (with a C:M ratio in the format X:1) this patch sounds quite thin. When compared to some of the later tones discussed here (with a C:M ratio in the format 1:X) this patch sounds quite dull and uninteresting.

Sine +5 to Sine and Sine +7 to Sine

OK. This is where things start to get metallic and my inability to explain fine sonic differences in words becomes even more noticeable. This is also where all of the patches will sound quite sharp in the top octaves of your keyboard.

With both of these patches, the sound is brighter than Sine to Sine and there is a distinct metallic edge to the sound. As would be expected, Sine +7 to Sine is brighter than Sine +5 to Sine, and has more of a metallic edge. However, for both patches the sound is cohesive: you cannot hear a separate sine wave and metallic element.

We are using a different type of carrier : modulator ratio here. For the first time, the modulator is pitched above the carrier. However, more significantly we are not using an 1:integer ratio:

- Sine +5 to Sine has the modulator pitched five semitones above the carrier—this gives a frequency equivalent to 1.33 x the carrier frequency which could be expressed as a C:M ratio of 3:4.

- Sine +7 to Sine has the modulator pitched seven semitones above the carrier—this gives a frequency equivalent to 1.5 x the carrier frequency which could be expressed as a C:M ratio of 2:3.

Another issue to note here: the pitch of both of these patches is determined by the modulator, not the carrier. The modulator will determine the pitch when the C:M ratio is less than 1:2. When the C:M ratio is equal to or above 1:2 (that is the modulator is at least twice the frequency of the carrier) then the carrier will determine the frequency.

Sine +12 to Sine

Sine +12 to Sine uses a modulator an octave above the carrier, this gives a C:M ratio of 1:2. As you would expect with a 1:integer ratio, the sound is not particularly metallic but more resembles a (slightly dull) square wave. If you look at Sort of Square you will see that these two patches have a very similar architecture.

Sine +16 to Sine

If you weren't convinced by Sine +5 to Sine and Sine +7 to Sine, this is where you can have no doubt about the ability for FM patches to have a metallic quality.

For this patch the modulator is pitched 16 semitones above the carrier: this gives a C:M ratio of 2:5.

Sine +17 to Sine

If you need more convincing about the metallic qualities of FM, then Sine +17 to Sine should do it.

With the modulator pitched at 17 semitones above the carrier, this simple FM stack has a C:M ratio of 3:8.

The reason I am stressing the metallic qualities at these frequencies is that with a large gap between the carrier and the modulator it becomes far easier to generate a metallic sounds. However the timbre of the metallic sound becomes more shrill and piercing. With these lower ratios a clearer, more usable sound can be obtained.

Sine +19 to Sine

With Sine +19 to Sine we have a C:M ratio of 1:3. As with all of the 1:integer ratios, the tone of this patch is free from overtones and has no metallic element to it. However, it is quite a sharp tone, which may sound a bit thinner than Sine +12 to Sine (the previous 1:integer sound).

Sine +21 to Sine

With the modulator pitched 21 semitones above the carrier, Sine +21 to Sine has a C:M ratio of 3:10.

To my mind this is the first bell-like tone, rather than metallic tone. You may disagree. Even if you agree that this patch's tone has a bell like quality, that does not necessarily mean it will be ideally suited for the creation of bell like patches.

Sine +24 to Sine

Hopefully you're getting the idea now about what to expect. Sine +24 to Sine has a modulator that is pitched two octaves above the carrier. This gives a C:M ratio of 1:4. As you would expect, the tone is bright but does not have the overtones that you associate with a metallic sound.

Sine +31 to Sine

Passing over the 1:5 ratio, Sine +31 to Sine which gives us a C:M ratio of 1:6. Like the 1:4 ratio (and the 1:5 ratio which could be achieved by tuning the modulator to 28 semitones above the carrier), this sound is progressively more piercing. You may also describe this as sharper, harsher or thinner.

To my ear, and again you may disagree, there are two elements to the sound from this patch—the sine wave element and a distortion (almost metallic) element.

Sine +36 to Sine and Sine +48 to Sine

When I said that 1:integer ratios don't have a metallic tone, I lied. As you can hear from these two patches, it is possible to hear a metallic tone in both of these. However, we have reached high C:M ratios here (1:8 and 1:16, respectively). Most significantly, the difference in sonic terms between these sounds and the sound that can be achieved when using ratios that lie between these two ratios is comparatively subtle given the intensity of the sound.

Sine +21 and Sine +33 to Sine

Sin +21 and Sine +33 to Sine is the first example in this group of a patch with parallel modulators. The two modulators are pitched:

- 21 semitones above the root (a C:M ratio of 3:10), and

- 33 semitones above the root (a C:M ratio of 3:20).

As you can see one modulator (operator three) is an octave above the other (operator two), and so has a frequency which is twice that of the other. While the sound of this patch is quite sharp and very bright, the two modulators work sympathetically to develop the tone of the patch.

Sin +21 andto Sine +21

Sine +21 andto Sine +21 uses parallel modulators and a cascade as follows:

- operator two modulates operator one

- operator one is also modulated by operator three

- operator three also modulates operator two—this last modulation means that the effect of operator two on operator one is to change its waveshape so operator one is not being modulated solely by a sine wave.

You can hear the effect of the cascade by right-clicking in the matrix where the oscillator two column intersects with the oscillator three row—this will "gray out" the number and will cut the feed from operator three to operator two. The effect of operator three on operator one will remain.

At lower ranges, this patch sounds warmer with the cascade in place. This is perhaps counter-intuitive. However, in the higher ranges the sound is distorted when the cascade is in place (as would probably be expected). For this reason it would probably be wise to use some key scaling to cut back operator three in the higher ranges.

Envelopes

All of the sounds so far have been produced without using envelopes—the note has been either on or off and the modulator has had a constant effect on the carrier after the key has been struck.

123

This gives a consistent, but harsh and uninteresting sound. The real magic of FM arises when one operator changes in level. Here are a few examples (ranging from bad to good) of how envelopes can be used when creating FM sounds.

Level Drop

This patch, Level Drop, illustrates the effect of the level of the modulator falling over time.

Operator one is modulated by operator two. The C:M ratio is 1:4 (the Coarse Pitch for operator one is set to zero and the Coarse Pitch for operator two is set to +24). The modulator is controlled by a volume envelope which opens at its maximum when the key is struck and then falls to zero over a period of about 10 seconds.

As the key is struck the modulator modulates the carrier to the maximum extent. This gives the same sound as was produced in Sine +24 to Sine. However over time the tone softens and tends to a sine wave. This is not surprising—the effect of the modulator drops over time. When the carrier is no longer modulated, all you will hear is a sine wave.

This mimics natural instruments which tend to a sine wave as they sustain and their overtones decay.

Pitch Drop

This patch, Pitch Drop, illustrates the effect of the pitch of the modulator falling over time. You will hear that its effect is more extreme (and less controllable) than the effect produced by Level Drop. The levels of both operators are stable for this patch.

As before, operator one is modulated by operator two. The Coarse Pitch for operator two is set to -48 giving an apparent C:M ratio of 16:1. However, operator two has its pitch modulated by a pitch envelope (check out the Rhino manual if you want to read more about this). The pitch envelope is very much like the level envelope used in level drop—the pitch drops over a period of about 10 seconds. I have also set the pitch envelope slider to the maximum.

When the note is struck, the modulator is pitched up to the maximum extent, so instead of being modulated by a low note as may be suggested by the modulator's Coarse Pitch, the carrier is modulated by a very high note—this gives a fairly bright sound.

The pitch of the modulator then falls over the next 10 seconds, finally reaching a point where the note is so low it ceases to have an effect on the carrier (and only the sine wave of the carrier will be heard). As you hold the note, you may well hear several notes descending in pitch as the C:M ratio relationship continues to shift.

Try playing a chord—the initial sound is really quite wacky.

You may find using pitch envelopes difficult if you are programming sounds to be used in a musical context. However, if you are programming sound effects, this sort of sound may be ideal.

Pitch and Level Drop

If that last example wasn't enough, Pitch and Level Drop illustrates the sound if the pitch and level of the modulator drop simultaneously. This patch has the same setup as Pitch Drop but adds a level envelope to the modulator too.

The sound is similar to Pitch Drop. However as the modulator has less effect on the carrier over time, the effect of the lower pitch of the modulator becomes less dramatic (for instance you don't get the "echo" or "triple falling" type effect just before the modulator ceases to have effect.

Sine +17 to Sine w/Env

Let's move on a step and look at two musical examples.

For this first example we will take Sine +17 to Sine that we used earlier and tweak the level envelope of the modulator. The modulator envelope has been set to open to its full extent immediately. It then decays over about four seconds with a curvature of about 25, so it looks like an exponential curve—it falls very quickly at the start of the decay but the rate of decline levels out: think of a piano's volume envelope and you will be getting close.

Without the envelope the patch sounds like a metallic noise. With the envelope it is transformed into a simple bell. The effect of the envelope is to modulate the carrier to the maximum extent for a very short period and then to remove this modulation source very smoothly.

However, as the carrier has a basic envelope, the sound can sustain indefinitely which detracts from some of the bell quality of the patch.

Sine +21 to Sine w/Env

Sine +21 to Sine w/Env takes Sine +21 to Sine we used earlier and adds an envelope to the modulator and an envelope to the carrier. The envelopes are very similar to the ones used for Sine +17 to Sine w/Env, but with one difference. In the modulator envelope a "hold" stage has been added so that after the attack has reached the maximum level, the envelope stays open at its maximum level for approximately 15ms before it begins its decay phase.

The sound of the patch with the envelopes added is classic 1980s DX bells.

The effect of adding the hold stage to the modulator's envelope is to give a bit more impact to the attack of the note. The effect is subtle, but is noticeable.

In the lower key ranges this patch feels to detuned, so what I have done is a bit of key scaling on the modulator to reduce its effect for the lower keys. This makes the sound less bright but it also makes the sound cleaner.

Slowing the Modulator Envelope

Before we move on I want to cover one last aspect: slowing the modulation envelope with reference to the carrier envelope.

There isn't an example patch here, but these techniques are used in Synth Brass which is described below under Building Usable FM Patches. As the patch name may indicate, the techniques can be used to create brass sounds.

If you create a simple FM stack (C:M = 1:1) and set the carrier with a fast attack and set the attack for the modulator to be slightly (and only slightly) slower—for instance 5ms slower—the effect of the delay will be imperceptible if the operator is listened to in isolation. However, the interaction between the two operators works in such a way that the "spit" in the attack of a brass note is created. The brass effect can be enhanced by adding a bit of feedback to the modulator.

If the envelope's attack is slowed further then the modulator's attack can be heard—the spit is lost along with the brass effect. If the modulator's envelope's attack is increased then the spit is also lost.

A brass tone can also be accomplished with a single operator by slowing the attack and adding a bit of feedback to the signal. This gives a more muted tone.

Building Usable FM Patches

To round off this chapter I am going to build some patches—you can find these patches in the patch bank under the heading First Patches. All of these patches are intended to illustrate:

- how you can bring the various elements together to create a sound which is usable in a musical context, and

- how to move from getting a sound at random and calling that a patch, to creating sounds by design that you need for a specific project.

As with all of the patches in this book, my aim is to show how these sounds can be created from a clean slate rather than to illustrate how to reverse engineer a patch. Many decisions I have taken when designing the sounds are a matter of taste—you many not agree with my choices. That is fine: you will need to make decisions based on what works for your particular track.

These patches are presented in a finished state. To explain how they are constructed, the first step will be to switch off the modulators and most of the carriers so that we can consider the individual ingredients.

Pulsing Square to Square

This first patch takes advantages of Rhino's flexible envelopes to create a pulsating rhythm. Other than that, the patch is quite straightforward using a simple FM stack of two square waves (with the modulator also having some feedback to brighten the sound). If you play more than one note with this patch, be careful to synchronize all of the note starts or the rhythm will get fairly complicated (and whatever you do, don't try to play arpeggios).

Take a look at the images of the two envelopes in Figure 7.9.

These envelopes create a quite simple rhythm. The first was created as a one bar rhythm with beats on the eighth notes. The first, four and seventh notes (the 1, 2, and 4 beats) had their volumes

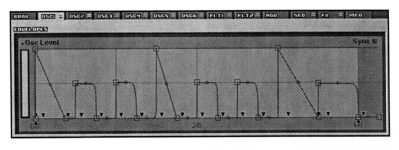

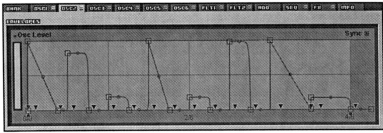

Figure 7.9   Creating rhythm with Rhino's envelopes.

emphasized. The decay on the emphasized beats was made more staccato than for the non-emphasized beats—for the non-emphasized beats, the gate time (the length for which the note sounds) was generally set to be around half of the note's length. This gives the basic pulsating rhythm.

This envelope was then copied to operator two and the level of each beat was then tweaked so that the tone shifts (slightly) as the rhythm plays.

The only velocity scaling that is used is in operator one. The effect is to control the volume of the patch with velocity. Velocity has no effect on the tone of the this patch.

... To Ruin

This patch is a take on the classic FM electric piano sound. You will hear that at lower volumes it is quite close to the classic sound, however at higher volumes it is brighter with more percussive bite.

For those of you who do not have the patches, you are going to need to set your envelopes to look like the image in Figure 7.10 (note that in the patch each envelope is individually set, for instance, operators five and six have decay times of around four and two seconds respectively).

Set the operators to the following coarse pitches: operator one: -12, operator two: -12, operator three: -12, operator four: 0, operator five: -12, operator six: +31. Note that all of the carriers (operators one, three and five) are tuned to the same pitch (-12).

Finally set the matrix with the following settings:

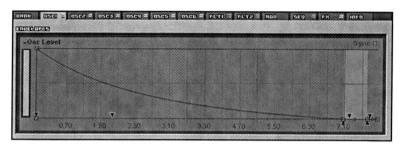

Figure 7.10   A classic piano type envelope in Rhino.

	Osc 1	Osc 2	Osc 3	Osc 4	Osc 5	Osc 6
Osc 1	021					
Osc 2	063					
Osc 3						
Osc 4			031	059		
Osc 5						
Osc 6			012		100	
Raw	100		067		045	

There are three elements to this patch:

- the "bell" sound, and

- the two parts of the sustain portion of the sound

   o   one part is more metallic, and

   o   the other more wooden.

Together the three elements all blend to form a cohesive sound.

Let's look at the elements individually.

First the bell tone. If you mute the outputs of the operators one and three, you will hear the bell tone on its own. This tone is created by operator six modulating operator five with a C:M ratio of (approximately) 1:11—I didn't use a scientific ratio here, I tuned operator six until it sounded right.

The output of the carrier is set to 45—this is the level that sounded right to my ear when mixed with the other carriers. However, the keyboard tracking (for operator five) has also been tweaked so the volume reduces slightly in a linear manner as the pitch of the note increases. Both operators five and six are fully responsive to velocity information, so not only does the volume of the bell change with velocity but also the quality of the tone changes, becoming brighter at higher velocities.

The bell element has a comparatively short decay time. To create the full sounds it is necessary to add the sustain portion. The tone I am after is quite complex and is very velocity responsive: at lower volumes it could be characterized as being more woody, at higher volumes more metallic. Accordingly there are two elements to the sound.

Turning to the metallic sound first, unmute operator three's output and mute the output from the bell tone. Now also mute the modulators of operator three and mute the feedback in operator four.

You will now have a sine wave which we want to convert into a metallic electric piano element. The basic tone is created by operator four modulating operator three with a C:M ratio of 1:2. Operator four is fully velocity sensitive so the tone is effected by velocity.

If you listen to this simple FM sound on its own, it doesn't sound that metallic—remember this is the sustained part of the metallic sound. To give the sound more bite and a metallic attack portion, operator six modulates operator three slightly—you will hear that the sustain section appears to sound more metallic when the new attack portion is added although there is no change to the sustain sound.

Finally, the feedback in operator four is increased. In isolation the feedback does not add a particularly pleasant quality to the sound. However, when the complete patch is played it does give some emphasis to the metallic element without changing the tone of the patch much.

Even at the lowest velocity levels I want some of this element to be present, so the lower end of the velocity scaling is set to around 20. In the context of the patch as a whole, I set the level of operator three to 67.

Lastly we will look at the wooden element which is created by operator two modulating operator one (C:M is 1:1) with a bit of feedback being added to operator one. Operator two is fully velocity sensitive. Like operator three, operator one has its velocity range slightly restricted.

The character of the two sustain elements when played together gives a broad range of tone from wooden to metallic. At higher velocity ranges they both take a more metallic tone. At lower ranges, the effect of the modulators is less pronounced and so the warmer tone can be heard.

Bringing all of the elements together gives a cohesive electric piano sound which does not have the feel of three elements glued together.

You may also notice that this patch uses one of the FX units: the rich chorus unit. This gives the patch a softer sound but with a bit more sparkle (and to my mind makes the sound slightly less digital).

You may note that operator four has been set to modulate operator one. This modulation is muted but is there as a further option. To my mind bringing in this element makes the tone too sharp. However, you may disagree or find a context where this tone may be appropriate.

As an alternative patch design, try the following. Mute output of operator six and mute all the modulators of operator one. This gives an alright but not overly interesting sound. Now increase the modulation effect that operator six has on operator one. Increase the level so that operator one is about 70% modulated by operator six. We now have a very usable four operator FM piano.

My opinion—and you may disagree—is that it is a fairly pointless exercise to create an FM electric piano patch from scratch: I have done it here to illustrate the general principles of FM patch design. There are many good patches out there (the internet is your friend). Do a search, audition some patches and then tweak your favorites to respond in a manner that is appropriate to your playing style. This is one areas where I think originality could lead to a lower quality result.

Organized

Organized creates an organ type sound by using five parallel carriers which are all modulated (to a greater or lesser extent) by the sixth operator.

For this patch, all of the envelopes are simple on/off affairs. For those of you without the patches, this is how to set up the sound:

	Osc 1	Osc 2	Osc 3	Osc 4	Osc 5	Osc 6
Coarse Pitch	-12	0	+12	+19	+24	-12
Fine Pitch	0	0	0	-19	0	-10
Modulation Amount by Oscillator 6	100	100	15	14	11	0
Output Level	100	100	100	27	35	0

If you are familiar with drawbar organs (or their software incarnations such as LinPlug's daOrgan or Native Instruments' B4) this patch will make a lot of sense to you. In this patch:

- operator one acts as the root

- operator two is the second harmonic (double the frequency of the root)

- operator three is the third harmonic (three times the frequency of the root)

- operator four is the fourth harmonic (four times the frequency of the root)

- operator five is the fifth harmonic (five times the frequency of the root), and

- operator six is the modulator for the other five operators.

Operator four has been detuned slightly to add a bit of movement to the sound: given the low level of this operator, the effect is quite subtle.

While operator one acts at the root, and operator two is tuned to the second harmonic, this relationship changes when the modulation is introduced. The effect of the modulator on operator two is that it sounds at the same pitch as the root. In practice, playing just operators one and two together (both with modulation) gives quite a rich tone.

Without the FX unit the sound is a bit intense. Switching in the FX makes the sound smoother and more organic—unsurprisingly, the rotary speaker simulator helps the patch to sound like an organ played through a rotary speaker. The chorus unit has also been added to soften some of the effects of detuning earlier in the sound chain.

Wind Instrument

Wind instrument is designed to recreate some of the tone of a wind instrument: it is not intended to replicate the precise characteristics of any specific instrument.

There are two main elements to this sound:

■ the breath noise—which is added in several places, and

■ the sustain tone.

Both elements of the sound are intended to be velocity sensitive—the breath noise in particular is intended to be noticeable at higher velocities.

As a first step, mute all of the outputs except for operator one. Also mute all of the modulations. You will now hear operator one alone. If you look at the envelope you will see that the operator has a fairly quick attack (about 3ms) and then a fairly quick (exponential) decay to its sustain level. For the sustain portion of the envelope you will see that the level varies slightly to give a tremolo type effect.

Operator three and operator five are both noise sources—white and brown noise respectively. If you look at the envelopes you will see that operator three (white noise) gives a pulse of noise—it fades in and out over approximately 5ms. By contrast, operator five has an envelope which resembles operator one's envelope (in broad terms). The two noise sources are used in different ways in this patch:

■ white noise (which is the brighter) exaggerates the breath noise at the start of a noise, where

■ brown noise is used to give coarseness to the note's tone to resemble the quality of a woodwind instrument.

Unmute operators three and five as modulators of operator one and unmute the output of operator three. Listen to the effect of the noise which at this stage will have a disproportionate effect on the sound of the patch.

Now mute the noise sources and unmute the modulation of operator one by operator two and unmute the feedback on operator two. You will hear a sustained woodwind sound—you will also hear that the level envelope gives the effect of breath at the start of a note. There are two envelopes affecting this sound: operator one's envelope which we have already discussed and operator two's envelope.

Operator two's envelope has an attack and sustain which is similar to operator one's. However, as can be seen in Figure 7.11, during the decay phase its level falls to zero and then increases again to the sustain level. In isolation this effect may be too subtle to hear. However, when listened to as part of the whole patch, the effect is to allow the sound of the noise sources to be heard more distinctly.

131

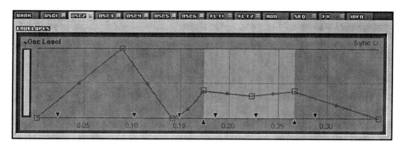

Figure 7.11   A volume envelope where the level of the operator subtly shifts during
the sustain phase.

This operator also has an element of velocity scaling and key scaling—this time gently cutting
the volume of the operator as the velocity increases and cutting the volume quite sharply above
note F5. Both of these scalings have been designed to allow the "chiff" of the woodwind to come
through more clearly.

OK, mute everything apart from the output to operator four. This operator is intended to thick-
en up the sound from operator one and so it is pitched 12 cents below that operator. It also has a
much more simple envelope: a slow attack and a constant sustain phase.

Operator four is modulated by two operators:

■   operator two which changes the tone of the operator to be more flute like (the attack envelope
    of operator two is largely ignored due to operator five's slower envelope), and

■   operator five which adds a bit of breath noise.

Operator four is fully velocity sensitive and has its level set at 60 which is where it sounded right
to me—it gave depth to the sound without creating the impression that there were two sound
sources.

Finally a very small amount of operator three (white noise) is also fed to the output just to add another shade to the breath noise.

Synth Brass

This patch takes a different approach to the other patches in this section in that it creates its sound
by a combination of FM and subtractive synthesis with two of the operators using a sawtooth wave
fed through a filter.

This patch is slightly counter-intuitive: the fatter element of the sound is created by the FM elements and the brighter elements of the sound are created by the subtractive synthesis. However,
this shouldn't really be that surprising: subtractive sounds are built around very bright sounds being filtered to remove the unpleasant high frequency element. FM sounds start dull and become
brighter when modulated.

Let's look at the subtractive elements first. Operators four and five both have sawtooth waves selected. These are slightly detuned and fed into filter one which has a low-pass filter selected (Analog Lowpass 2). You will not hear the raw output of these oscillators in this patch.

Each of the operators is controlled by an envelope. The attack time is fast (but not zero, so there is a brass quality to the attack). The envelopes then enter a short hold stage so the envelope will be fully open for a few milliseconds. This hold stage gives the sound a bit more punch when compared to an envelope where the decay phase begins immediately. The envelope then drops to, and remains at, the sustain level. There is no velocity or key scaling applied to either of these operators.

The filter cut-off frequency is also controlled by an envelope. It also has a fast (but not zero) attack time, but there is no hold phase. The filter is velocity sensitive so it opens up more at higher velocities.

The sound so far is acceptable, but not particularly impressive. As I have mentioned earlier in the book, adding a sine wave to a patch is a simple way to add weight. Operator six does this without detracting from the essential tone of the patch.

Its time to move on to the FM elements, so mute operators four, five and six.

I am going to create an FM brass sound which will fatten up the sound when added to programming we have already completed. As I want this sound to have a fattening role, it is crucial that it doesn't have too bright a tone.

Brass sounds can be created with a simple FM stack using a C:M ratio of 1:1. The key to the brass tone is ensuring that the modulator's attack envelope is slightly slower than the carrier's attack envelope. In this patch we are going to use a parallel carrier arrangement to create the brass sound.

Operators one and three (the carriers) are set with a very basic envelope—it is essentially on/off but with the attack phase is slowed slightly to give a brass like character. Both carriers are fully velocity responsive. To add some movement to the sound, carrier three is detuned slightly (-3 cents).

Operator two acts as the modulator. Its level envelope has a slightly slower attack than either of the carriers' envelopes. It also has a hold phase which, in conjunction with the slowed attack, gives some emphasis to the brass attack sound. After this the envelope falls to the sustain level. This modulator is largely (but not entirely) velocity sensitive to allow for tonal variation to be controlled by velocity.

To give a tonal variation between the carriers, different amounts of modulation are applied—operator one's modulation is set to 100 and operator three's to 59. To give the modulation a bit more brightness a very small amount of feedback is added to the modulator. To further change the tone of operator three, an amount of feedback is added.

Once this FM section has been set up, add the other three operators and travel back to the 1980s.

Clangy Bell

As you will have realized, FM synths are great for metallic sounds. However, controlled and usable FM sounds take some work. For this patch I wanted to create a bell sound, but not one that is too clean: I am interested in a sound that has a good clang to it and can be used over a wide range of the keyboard. I am not concerned about the velocity scaling or the nuances of the bell, nor whether it can be played as a chord.

In essence, this patch is going to be built around two simple FM stacks creating a fairly deep bell and a third simple FM stack giving a brighter overtone.

First let's look at operator six which modulates carriers one, three and five. This modulator is working as a low frequency oscillator using a triangle wave. However, instead of having a fixed frequency for the LFO, I have set it with a C:M ratio of 3:1 (by setting its Coarse Pitch at -19 where the carrier's Coarse Pitch remains at 0). In this way the rate of the LFO is dependent upon the pitch of the note. This will ensure that if two notes ring the LFO effects do not work in sympathy (which really wouldn't sound natural for a clanging bell). Now mute the LFO modulations until we have finished building the patch. Apart from the LFO, all of the level curves follow the familiar pattern of rapid decay that we have used earlier.

Operator one and operator five work as parallel carriers—to give their bell-like tone, they are both modulated by operator two which is pitched at a 1:9 C:M ratio (having its Coarse Pitch set at +21). This tuning was determined by ear as was the feedback which is added to operator two to give some brightness. Operator five was then slightly detuned. If you engage only operators one and five and only modulate them with operator two you will hear the volume cancellation of this effect. This effect will be counteracted with:

- a further change to the tone of operator five (when it is modulated slightly by operator four)

- the LFO, and

- the sound output from operator three which disguises some of the volume fluctuation.

Operator three is intended to add some more brightness, but given that the patch is intended to be a "clangy bell" (rather than a crisp, clean, clinically sharp bell) it is important that this operator doesn't get too bright and so the ratios are kept low.

For now, mute the other two carriers and unmute operator three (and mute its modulators if you haven't done so already).

Operator three is primarily modulated by operator four. To ensure that the sound does not become too shrill I selected a low C:M ratio—in this case 1:2 (by tuning this operator up one octave) and added a reasonable amount of feedback to the modulator to dirty up the signal a bit. To also add a bit more bite, a small amount of operator two has also been fed in—this gives a touch more bell-like attack to the sound.

On its own the sound of operator three isn't particularly pleasing now: with the modulation it gives a buzzy almost distorted sound. However, when you add the bell sounds back (and the

LFO), the sound is much more pleasing and realistic with a far more cohesive tone than may be expected if you listen to the elements individually.

The last task before this patch is finished is to sort the key tracking. The patch benefits from some tweaking of operators four and five. Operator four has its effect cut (quite drastically) below octave five, and operator five is cut equally drastically above octave seven. Without these changes the sound in the lower key ranges had too many overtones which to my mind made the bell indistinct and unfocussed, rather than clangy. Without the change in the upper key range the sound had an unpleasant, fizzy attack.

Another Bell

While Clangy Bell has a great sound it may not always fit with a track. First, it does not have a modern pop type sound and so perhaps it would be more useful as a special effect and secondly, it takes up a lot of the sonic range, in other words it could drown out other important aspects of your production.

If you can't use Clangy Bell, then you may need Another Bell. This is a much brighter and more modern bell sound (although its heritage is something of a 1980s DX cliché which may discourage its use) and you can play chords without too much dissonance.

The difficulty when trying to construct a bell is that the very essence of the bell sound is the many different frequency components which are not necessarily harmonically related. Without these elements a bell is just a high pitched noise, but with too many of these elements the sound becomes difficult to pitch. While FM can readily create the elements, it is important to balance them so that the sound is useable but retains that bell like quality.

The foundation for Another Bell is four parallel carriers (operators one, three, four and five), three of which are modulated by one modulator (operator two). The four carriers are all pitched at the same level, although operators four and five are pitched 32 cents below and 16 cents above the pitch to give a bit more ring to the bell (and introduce a bit of vibrato and tremolo).

The envelopes all follow the form used in Clangy Bell—for operators one, four and five the decay time is around seven seconds. For operator two (the modulator) the decay time is around one second, and for operator three the decay time is just over a second.

Operators one, four and five are all modulated by operator two. The amount of modulation varies to change the tonal quality of each of the carriers: the respective modulation amounts are 47, 74, and 100. The modulator is pitched to a level where it sounded right so there was enough bell sound without there being too many overtones which would have made pitching the sound difficult. In this case the C:M ratio is 1:7, in other words its pitch is raised by 34 semitones.

Operator three has been added to round out the sound during the attack phase. While subtle, it does give the sound a slightly fuller tone.

A further subtle change to vary two of the three main tonal elements is the additional modulation of operators one and five (by operator three). As I said, this is subtle, but it does add a slightly

135

brighter quality to the attack of the note (which is balanced with the roundness added by operator three's raw output).

Operators one, four and five have been made fully velocity sensitive. At quieter volumes the effect of operator three, which has no velocity scaling, becomes more apparent. The modulator, operator two, has also been given an element of velocity scaling to reduce some (but not all) of the brightness of the tone at lower levels.

Finally, at higher key ranges (above octave six), the effect of the modulator has been reduced in order to tone down some of the harshness that would otherwise be apparent in the note.

Bass

The skip from a bell to a bass sound may seem odd, however, the foundation of Bass is the bell element—this is what gives the sound its bright attack. Bass takes the next step forward after FM Bass which was constructed at the start of this chapter.

The challenge with any bass sound is to capture the bass element (obviously) of the sound and the brightness of the sound without these two elements being heard as two distinct sounds—in a bass patch, you should hear a cohesive sound.

So how do you get these two elements in one sound? Simple (almost)—use parallel modulators.

Open up Bass and mute all of the modulations and all of the raw outputs apart from operator one's output. You will hear a plain sine wave with a decay of about 7 seconds. Now unmute the modulation of operator one by operator three. Operator three is pitched an octave below operator one and has the effect of giving a bass tone—there are many circumstances where this simple FM two operator sound alone would be sufficient as a bass sound. However, our goal is a fuller, touch sensitive, bass sound.

You will also hear that operator three has a much faster decay envelope than operator one. Operator three's decay lasts for just over a second so you hear the bass tone initially and as the note sustains you hear a much higher sine tone. You could extend the decay time of the modulator to give a longer deep tone. However, I don't want to for several reasons:

- First, I don't want the bass sound to take up too much of the lower frequency spectrum—with this patch design the bass element is heard when the note is struck. This design makes the bass have more impact but without making it staccato and so it should fit into a mix more smoothly.

- Also, to my ear the tone sounds duller if the envelope for this modulator is extended. As always, you may disagree with this subjective assessment.

Now mute the modulation of operator one by operator three and unmute the modulation by operator two. Immediately you will hear the bell tone I mentioned earlier (we are using a C:M ratio of 1:7—this is the same ratio that we used in Another Bell) with operator two's pitch raised by 34 semitones. This is not the greatest bell tone we could design, however, it does give some brightness to the sound.

Unmute operator three so we have the parallel modulator arrangement in place. Now we have achieved our goal—a bass sound with some brightness which retains its cohesion. However it still needs a lot of work.

On a side note, it would have been possible to pitch operator three an octave lower. This would have given an even more "bassier" sound. However, this would have led to two undesirable side effects:

- first, the two elements of the sound are heard separately so the whole cohesion of the sound that I've been trying to achieve would be lost, and

- second, the bass starts to sound flabby.

The first thing I want to add to the sound is an additional bell element to keep the brightness of the attack. To ensure the patch remains as a bass patch and there are no weird metallic overtones, I have duplicated the existing bell element (operator one modulated by operator two) and pitched it an octave lower.

This lower bell element is created with operators four and five. Operator four is tuned to an octave below operator one. Operator five is then tuned to be an octave below operator two—this maintains the C:M ratio of 1:7. Operator four is then very slightly detuned to take off some of the sharpness of the duplicated bell sound.

The sound needs some rounding out—it is after all intended as a bass sound. To do this, I brought in operator three as a raw feed. You will remember that operator three is pitched an octave below operator one and already modulates operator one to create the bass tone. Earlier I considered increasing the time of operator three's decay envelope: if I had done that it would have had a negative impact here too.

Now that the key elements are in place we can add some scaling. The only key scaling I want to add is to operator two so that its effect is reduced at higher key ranges (above octave five). A similar process could be applied to operator five above the sixth octave, however, as that range is getting out of the usual range of play, I am not too concerned to make the change.

I did want to add some velocity scaling to control both the volume and the tone of the patch. To do this I made all of the operators velocity sensitive to a greater or lesser extent. The only place where I did something out of the ordinary is with operator two where the velocity scaling is not linear. Instead the velocity scaling is curved so the effect of the operator becomes much more apparent at higher velocities.

You will hear that this patch can be played in higher ranges. However, if I was going use this patch in this way, I would want to tweak it a bit to get a more even response across the keyboard.

FM Patches: What Did We Do?

It is not the style of this book to summarize the content of the chapter that you have just read. However, it may be useful to draw together some of the main design ideas that were touched upon in these sound examples:

- A C:M ratio of 2:1 (or even 3:1) will generate a very usable bass tone which can be used as a foundation for many patches.

- Parallel modulators can create a cohesive sound that has both a deep tone and a bell-like sparkle.

- Sine waves are great for rounding out sounds.

- High C:M ratios (where the modulator is pitched above the carrier) are where the metallic sounds are found.

- Sounds from 1:integer C:M ratios produce cleaner tones.

- Increase modulation for brighter sounds and cascade modulation for more detailed and complex sounds.

- The pitch of a patch can be affected by C:M ratios.

- Fixing the level relationships in cascades (giving less velocity sensitivity) can be good if you are after a specific tone.

- Envelopes need attention—in particular, think about:

  o fast decay time to give emphasis to the transients at the start of the note

  o the attack "spit" with brass sounds, and

  o introducing a hold phase for emphasis.

FM Patches—What's Next?

While we have built a fair number of FM patches in this chapter, this isn't the end of the FM examples in this book. As I've mentioned several times, check out the supplement to this chapter which revisits the whole topic but uses Surge which has some great FM features including the facility to use up to 24 operators.

In addition, chapter 8: Wave-Sequencing gives many more examples of FM patches. In particular, that chapter introduces the concept of FM wave-sequencing. You should also check out Chapter 9: Additive Synthesis, which features more examples of patches built with Rhino but this time without using FM techniques.

There are also some more FM patches (using Rhino, Surge, and Z3TA+) set out in Chapter 12: Patch Building. You will also find some ring modulation patches build with Surge in Refill#2.

Chapter 7 Supplement

Frequency Modulation Synthesis with Surge

Chapter 7 gives a comprehensive introduction to frequency modulation synthesis. I suggest you acquaint yourself with that chapter before you dive into this supplement which, as you might have guessed, deals exclusively with creating frequency modulation sounds in Surge.

Surge is effectively two synthesizers bolted together: each synthesizer (or Scene as they are called) operates independently. The outputs from the synthesizers then share some FX and are mixed before they reach the output. Each Scene has three oscillators. As Figure S7.1 shows, when creating FM sounds, these three oscillators can be arranged as:

- A simple FM stack with the third oscillator operating independently.

- A parallel modulator arrangement with oscillators two and three modulating oscillator one.

- A cascade with oscillator three modulating oscillator two, which then frequency modulates oscillator one.

Figure S7.1 The FM algorithm selector in Surge. In this instance, the parallel modulator option has been selected.

With these three algorithms, the modulators can also feed their signal directly to the output—they are therefore not limited to acting solely as a modulator. There is also the added flexibility that each Scene can have a different algorithm (and indeed, one Scene could use FM and the other not).

Each oscillator can use any of Surge's waveforms—this gives a great range of sonic options but it can have variable results when creating FM sounds. There is one minor lack of flexibility where the

FM Depth (in other words, the FM index—the amount by which the modulator(s) modulate(s) the carrier) is the same for both modulators.

However, the real power for Surge comes with the specialist FM oscillators which were introduced with version 1.5 of Surge. These specialist oscillators give you the following options.

■ **FM2.** The FM2 algorithm (see Figure S7.2) has two parallel modulators, each of which can have its pitch and FM index set individually. The algorithm also allows these modulators to be detuned and for the phase of the modulators to be adjusted. Finally, the unit allows for feedback to be introduced in the carrier.

■ **FM3.** The FM3 algorithm (see Figure S7.3) takes a step further than FM2 and offers three parallel modulators, each of which can have its pitch and FM index set individually. You can also add feedback to the carrier.

Figure S7.2   The FM2 oscillator in Surge.

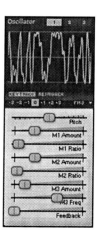

Figure S7.3   The FM3 oscillator in Surge.

These new oscillators offer many sound generating possibilities, and are highly flexible, especially when combined with other features, such as the Classic oscillator and the filters. By using all six of Surge's oscillators, you can create detailed and complex sounds using up to 24 operators. However, there are a few restrictions on what you can do with these algorithms.

■ There is no feedback option for any of the modulators (but there is for the carriers).

■ There is no direct audio out for the modulators—they act solely as modulators.

■ The modulators cannot be pitched below the carriers. If you want to do this, then you can still use the individual oscillators and the FM Routing options (but then you can't use the dedicated single oscillator FM algorithms).

■ There are no parallel carrier algorithms.

As you will hear, Surge's strength with FM isn't so much with the classic (or if you prefer, cliché) FM sounds. Where it really excels is in providing a really good range of usable and controllable

FM tones. Think of it as FM without the craziness and you'll be getting there. In short, it's a really useful tool if you want to create FM sounds.

I'm now going to move on and create some sounds with Surge. Many of these sounds are the counterparts to the sounds I created in Chapter 7 and so you will see that I have not discussed the sounds in detail. However, as you can hear if you compare the two, you will notice that there are significant sonic differences between the two synthesizers. I don't think the differences are good or bad—instead, the differences give an even broader sonic palette.

On the whole, I feel that Surge's sounds tend to be fuller/thicker/brighter. However, Rhino offers more control and routing options allowing for more delicate tones in certain circumstances. This is just my perspective on the issue—you're going to have to make up your mind about which synthesizer is right for any given situation.

Building a First FM Patch in Surge

Before I go any further, I want to construct a basic sound using FM in Surge. 01 FM Bass is similar in character to its counterpart in chapter 7 that was constructed in Rhino, but this version sticks with three operators and doesn't have the tonal options of the version in Rhino.

As Figure S7.4 shows, the heart of the FM Bass patch in Rhino is a parallel modulator arrangement.

Figure S7.4 The routings in FM Bass when created in Rhino.

You will notice that in Figure S7.4, oscillator three modulates oscillator one and oscillator two as well as feeding its output directly to the audio stream. Unfortunately, we cannot set that routing so easily in Surge—oscillator three can modulate either oscillator one or oscillator two, not both. Instead, we will have to use two additional oscillators to establish this routing. We can't use one of the specialist FM oscillators since the modulator is pitched below the carrier so I've set up this routing with two further oscillators in Scene B. Figure S7.5 shows how the 01 FM Bass routings have been set up in Surge.

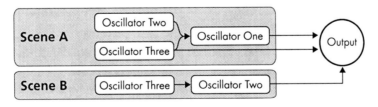

Figure S7.5   The routings in Surge for 01 FM Bass.

Creating this patch in Surge is quite straightforward. First I loaded a sine wave into the three oscillators in Scene A, selected the 2>1<3 routing option (so oscillators two and three work as parallel modulators), and then dropped the pitch of oscillator two by an octave and the pitch of oscillator three by 19 semitones (an octave plus a fifth).

Before I created the FM routings, I tweaked the Amp EG (the volume envelope). In this case, I kept the Attack time at zero, pushed the Decay time to around two seconds, and set the Sustain level at zero. Once the volume envelope was set, I also used this as the envelope as the modulation source to control the FM Depth. The FM Depth control was set at its minimum (-48 dB), and the Amp EG modulation was set to give around 13 dB of increase.

To create the tone from oscillator three modulating oscillator two, I copied Scene A to Scene B, and set the Scene selector to Dual mode. Although I've shown oscillator three modulating oscillator two in Figure S7.5, if you open the accompanying patch, you will see that in fact, oscillator two modulates oscillator one. This is a necessity due to the routing options. I've retained the labelling in Figure S7.5 to reflect the pitching of the oscillators to make a comparison with Figure S7.4.

To fit with the routing, I chose the 2>1 routing option and to follow the pitching of the oscillators in Scene A, I set the pitch of oscillator one (which doubles oscillator two in Scene A) to -1 octave, and set the pitch of oscillator two (which doubles oscillator three in Scene A) to -19 semitones. To tweak the tone, I then pushed the FM Depth slider to -40 dB. FM Depth is still modulated by the Amp EG (as this is part of the copied Scene) and the relative modulation amount remains.

If you compare this patch to its counterpart in Rhino, I think you will find this one to be fatter and warmer. This may, or may not, be useful depending on the context of the track in which you are intending to use the sound.

You can also see that while this patch might appear to be more complicated, the FM routing has been quick to set up—as there is only one FM Depth control and only one envelope, I have had to spend a lot less time tweaking the patch.

Building Blocks of FM Sound

I now want to walk you through the basic elements of the FM sound but this time using Surge. I'm aiming to achieve two goals with this:

- First, to introduce you to the basic range of FM sounds. This is especially relevant if you haven't been through these sounds in Rhino.

- Second, to allow you to compare and contrast the FM sounds created in Rhino and Surge. To my mind, the individual sounds in Surge have a wider range of tones than the comparators in Rhino. While this may seem like a good thing—and it often is—do remember that there will be times when you are creating sounds when you will want less sonic flexibility.

1:1 Ratio—FM Using Different Modulators

This first group of patches illustrate the basic tones that are available in a simple FM stack by modulating a sine wave with various different waveforms. I have used five different waves although Surge has many more waveforms to choose from.

For this group of patches, the modulator is touch sensitive. This will give you an idea of the effect of the level of the modulator on the tone of the patch. You should also note that for these patches the envelopes are very basic (the operator levels do not change over time).

For those of you without the patches, you can set up Surge for this group of sounds in the following way:

- Oscillator one (to act as the carrier)—octave pitch set to 0.

- Oscillator two (to act as the modulator)—octave pitch set to 0.

- Amp EG—Attack time set to 0, Sustain level set to full.

- Oscillator FM Routing—select 2>1.

- The modulator is fully velocity sensitive. To set this, set the FM Depth to zero and then click twice on the Velocity selector in the Route section (the blue strip of boxes above the LFO section). Once in modulation assign mode (the slider caps will turn blue), set the FM Depth slider to the maximum and then click on the Velocity selector again.

The patches in this section are:

- 02 Sine to Sine

- 03 Triangle to Sine

- 04 Square to Sine

- 05 Saw to Sine

- 06 Noise to Sine

The waveforms used are suggested by the patch names.

While the sonic results are different from Rhino, many of the characteristics of the patches are similar, so I'll leave you to look back at the descriptions in Chapter 7. However, as you can hear if you compare Rhino and Surge, in Rhino some of the patches exhibited a tendency for two sounds to be heard (for instance, you could hear a sine wave and a square wave when the square wave frequency modulates the sine wave). This is not the case in Surge where a clear, cohesive, frequency modulated sound is created.

1:1 Ratio—FM Using Different Carriers

The next group of patches illustrate the basic tones that are available in a simple FM stack when different waves are modulated with a sine wave. For these examples the modulator in each of the patches is again touch sensitive. This will give you an idea of the effect of the level of the modulator on the tone of the patch.

As before, the waveforms are suggested by the patch names, and I'll leave you to look at Chapter 7 if you want a fuller explanation about the sounds. The sounds in this section are:

- 07 Sine to Triangle

- 08 Sine to Square

- 09 Sine to Saw

- 10 Sine to Noise

Without wishing to make crass generalizations, with Surge I feel that when the sine wave acts as a modulator (where the carrier isn't a sine wave), the range of tones that is available is much narrower than when another wave modulates a sine wave (or a sine wave modulates a sine wave).

1:1 Ratio—FM Using Same Modulator and Carrier

This last group of patches illustrates the tones that are available in a simple FM stack when an operator is modulated by the same waveform. For these examples the modulator in each of the patches is again touch sensitive. This will give you an idea of the effect of the level of the modulator on the tone of the patch. As before, the envelopes are very basic (the operator levels do not change over time).

The patches in this group are:

- 11 Triangle to Triangle

- 12 Square to Square

- 13 Saw to Saw

- 14 Noise to Noise

As with the last group of patches where different carrier were modulated by a sine wave, this group of patches shows much less tonal variation when compared to a sine to sine set up, or the patches where different waves modulated a sine wave.

1:1 Ratio—FM Using Cascading Operators

I only want to look at one example of a cascading operator set up: 15 Sine to Sine to Sine. With this patch, you can hear that there is a huge tonal variation. You can also hear that as the respective levels of the modulators change (with varying velocity input levels) that not only does the tone change, but the pitch of the sound also appears to change considerably.

As I mentioned earlier, unlike Rhino, with Surge you cannot fix the FM Depth for one of the points in a cascade. This may restrict some of your tone shaping options.

Varying Carrier : Modulator Ratio

The sounds in this next group are all created with a simple FM stack where the modulator is at a different pitch by reference to the carrier. I should point out that the carrier/modulator pitch intervals illustrated here are not the only options, and there is no reason for you to stick with the semitone intervals that I have used. I should also mention that unlike the Rhino comparators, the modulators in this group of patches are all touch sensitive.

In the patch names I have described the interval between the modulator and the carrier in terms of semitone steps. Check back to Chapter 7 if you want to see a fuller explanation of the frequency ratios.

The patches in this group are:

- 16 Sine -36 to Sine
- 17 Sine -24 to Sine
- 18 Sine -19 to Sine
- 19 Sine -12 to Sine
- 20 Sine +5 to Sine
- 21 Sine +7 to Sine
- 22 Sine +12 to Sine
- 23 Sine + 16 to Sine
- 24 Sine +17 to Sine
- 25 Sine +19 to Sine
- 26 Sine +21 to Sine
- 27 Sine +24 to Sine
- 28 Sine +31 to Sine
- 29 Sine +36 to Sine
- 30 Sine +7 and Sine +19 to Sine

This last patch is a parallel modulator arrangement where one modulator is pitched an octave above the other. As you might expect, this has wide tonal variations, and since there are two modulators, it also has the facility to tend to noise.

Envelopes

All of the sounds so far have been produced without using envelopes—the note has been either on or off and the modulator has had a constant effect on the carrier after the key has been struck. This gives a consistent, but harsh and uninteresting sound. The real magic of FM arises when one operator changes in level. Here are two examples of how envelopes can be used when creating FM sounds.

31 Level Drop

With 31 Level Drop, you can hear the effect of the level of the modulator falling over time.

Oscillator one is modulated by oscillator two. The C:M ratio is 1:8 (the oscillator pitch for operator one is set to zero and the oscillator pitch for operator two is set to +3 octaves). The modulator is controlled by the Filter EG which decays over a period of about 10 seconds.

As the key is struck the modulator modulates the carrier to the maximum extent. This gives the same sound as is produced when 29 Sine +36 to Sine is struck with maximum velocity. However over time the tone softens and tends to a sine wave. This is unsurprising—the effect of the modulator drops over time. When the carrier is no longer modulated, all you will hear is a sine wave.

This mimics natural instruments which tend to a sine wave as they sustain and their overtones decay.

32 Sine +17 to Sine w Env

32 Sine +17 to Sine w Env takes 24 Sine +17 to Sine that we used earlier and replaces the velocity sensitivity with an envelope to control the FM Depth. The envelope used in this case is the Filter EG (which has no effect on the filter in this patch since none of the filters is used).

Without the envelope the patch gives a range of metallic bell-like tones. With the envelope it gives a much more consistent, plucked metallic sound. You can tweak the envelope (and the amount of modulation in the FM Depth) to change the tone.

Building Usable FM Patches

To round off this supplement, I want to build only one patch: 33 Pulsing. This is another take on the Pulsing Square to Square patch we created in Rhino.

It is a very simple patch, and only uses on oscillator, albeit I am using the FM2 algorithm oscillator.

With this patch, modulator M1 has its pitch (M1 Ratio) set at 7 and modulator M2 has its pitch (M2 Ratio) set at 3. I have then used a similar rhythmic pattern to that used in Pulsing Square to Square. However, for this patch, I have created this rhythm with the Step Generator in Surge. LFO 1 creates the main pattern and modulates M1 Amount, and LFO 2 creates a more simple pattern to emphasize the beats, and modulates M2 Amount.

Again you can see and hear that this give a more up-front tone, but is much quicker to construct than the Rhino equivalent.

Chapter 8

Wave-Sequencing

As I mentioned earlier, in essence, wave-sequencing is the process of stepping through a series of waves in a predetermined order. This allows you to:

■ create rhythmic patterns by varying the waveform, and/or

■ shift the tone constantly as a note is sustained.

Wave-sequencing is not a different way to create a sound source in the same way that FM is different from subtractive synthesis. However, with a constantly shifting tone, you can create sounds which cannot be created with either subtractive synthesis or FM. For instance you can:

■ cross-fade between waves to get an effect close to morphing, and

■ create "infinite" waves, for instance you could line up 30 waves each subtly different so that the sound changes gradually over time, giving you a highly unique oscillator.

These two effects can be created using additive synthesis, however, that technique requires far more work to create the result which can be quickly achieved using wave-sequencing.

Basic Wave-Sequencing

It may be as impractical to teach someone how to "wave-sequence" as it is to teach them how to write a song—at the end of the day, it all comes down to knowing the basic techniques and a matter of taste. However, a few very simple wave-sequences may help to illustrate the power of the technique and some of the difficulties with sequences.

The most obvious synthesizer to use to demonstrate wave-sequencing is Wusikstation—it was designed to wave-sequence. However, I first want to demonstrate the technique with Rhino and Surge. Hopefully this slightly more "hand cranked" demonstration will make it easier to understand what is going on. As you would expect, you will find more sophisticated wave-sequences in the patches we build with Wusikstation.

For all of these patches, I suggest you set the tempo of your host to around 110bpm. While this first group of patches uses Rhino, none of them uses any FM.

Sinsqutrisaw

This first patch is a four beat wave sequence:

- on the first beat a sine wave sounds
- on the second beat a square wave sounds
- on the third beat a triangle wave sounds, and
- on the fourth beat a sawtooth wave sounds.

The construction of the patch is quite straightforward—the different waves are allocated to the four oscillators. To create the basic rhythm, the Basic 4 Env envelope in the Rhythms folder was selected and its length stretched by 200%. This causes all of the oscillators to sound together on the beats.

To get the sequence to play, you can edit the envelopes by dropping the volume peaks where the wave is not wanted so that the peaks that are left play the waves in the correct sequence. In other words, the envelope in oscillator one peaks on beat one, the envelope in oscillator two peaks on beat two and so on.

This first sequence illustrates much of what is interesting with wave-sequences and much of what is disappointing with sequences. On the plus side, four simple waves have been taken and an interesting rhythmic sound has been created. However, on the negative side, the sound soon becomes quite repetitive.

It is also interesting that the comparatively weak sine and triangle waves seem to fit the sequence better than the unfiltered sawtooth wave which comes across as being slightly buzzy to my ear.

Sinsqutrisaw Fade

This second patch is a copy of the first, but with a few edits made to the envelopes. First the length of the envelopes is increased by 200% and secondly I have adjusted the peaks so that instead of each wave sounding separately, they cross-fade. While this gives the sound of a shifting tone rather than four distinct waves, you can clearly hear each separate wave—by no stretch of the imagination is this one constantly shifting tone.

It is also interesting that in this patch, the square wave and the sawtooth wave both seem far more dominant than they were in the more rhythmic patch.

Sinsqutrisaw 8ths

So far the sequences have been fairly linear, playing one wave after another. With this next patch we will again take the first patch and will edit it to do two things differently:

- first, the order of the waves will not be sequential, and
- second, more than one wave may play at once.

To liven things up a bit, I loaded the Basic 8 envelope for each of the four oscillators and I have set the levels as described in the Table 8.1.

Table 8.1   The level settings in the Sinsqutrisaw 8ths patch.

	Step One	Step Two	Step Three	Step Four	Step Five	Step Six	Step Seven	Step Eight
Oscillator One: Sine	Full Volume			Full Volume		Full Volume	Full Volume	Full Volume
Oscillator Two: Square		Full Volume	Half Volume	Half Volume	Full Volume			Full Volume
Oscillator Three: Triangle			Full Volume				Full Volume	Full Volume
Oscillator Four: Saw	Full Volume		Half Volume	Full Volume	Full Volume		Full Volume	

As you can hear this gives a rhythm that is far more interesting than that in the first patch.

Additive Seq

For Additive Seq, I have taken six sine waves. Waves two to six are pitched a frequencies which are the multiples of the first. As a side note, these sine waves have been created using Rhino's additive wave generator—the basis of this patch will be used again in the next chapter (Additive Synthesis).

Once the waves are set up, this sequence then creates an arpeggio pattern from the waves.

The prime purpose of this patch is to illustrates my frustration with melody auto-play patches. While the effect is interesting, the result is highly limiting—for instance you can't play chords and (in this case) there isn't a minor arpeggio.

Additive Seq II

Moving on, Additive Seq II takes the foundation of the previous patch but creates a more rhythmic rather than a melodic sequence. This latter patch uses the same sine waves at the same pitch as are used in Additive Seq. However, the higher pitched waves are largely used as harmonics shaping the tone rather than as a fundamental note and so simple chords are possible with this patch.

The rhythm was created in a similar way to how the rhythm in Sinsqutrisaw 8ths was constructed. Those of you with the patches will see that with oscillators five and six some notes are held so they are not just a rhythmic beat, and that some of the beats fade in (giving a wine glass-type bell tone).

For wave-sequence patches which create rhythmic patterns, the MIDI part playing the synthesizer must be simple: programming the patch is more akin to programming another rhythm part. You are going to need to think how the wave-sequence will work with the other rhythmic elements (primarily drums and bass) when you are programming.

For tone shifting wave-sequence patches, players can often adopt a conventional playing technique. However, this style of patch is often more suited to sustained chords and so perhaps some aftertouch could introduce some interesting modulation effects.

A Different Approach to Wave-Sequencing

Surge allows you to create the basic wave-sequences that you can create in Rhino. With the Step Generator, these sequences can be put together much more quickly than they can in Rhino, however, some of the detailed editing that can be introduced with Rhino's flexible envelopes is lost in Surge.

Surge also offers another option for creating wave-sequences. With the wavetable oscillator, instead of simply giving a range of single oscillators, Surge offers a number (up to 1,024) of grouped, related single-cycle waveforms. You select the oscillator group and the Shape parameter allows you to choose the specific waveform. If you then automate the Shape control (by controlling it with a modulation source such as the Step Generator or one of the LFO waveforms), then you can create a wave-sequence.

The wave families have a certain relationship and so there are sonic similarities. This means that if you play two adjacent waves, you can get a tone shift which is similar to a filter closing or a natural instrument decaying. While this can give a natural sound, sometimes you want a more extreme tonal shift in order to create a more rhythmic effect: this can be achieved by selecting a wave which is not immediately adjacent.

I will demonstrate some of the wave-sequencing features of Surge in the next three patches.

01 Sine WaveSeq

With this first patch, I selected the Sine wave set. I pushed the Shape slider until it was set at about 58% which gave me a really bright starting point.

From this position, I assigned LFO 1 to modulate the Shape slider to the maximum extent (1.00). In LFO 1, having chosen the sine wave, I set:

- Rate to 0.47 Hz so the sweep through the waves can be clearly heard.

- Phase/Shuffle to 25% so the maximum modulation is achieved when the key is hit.

- Magnitude to around 48% so that the sweep is only through a limited number of waves in the middle of the wavetable. In particular, this excludes the more obvious waves at the start of the wave table.

And that's it. As always with Surge, it is very quick to create sounds.

02 Cosine Octaves

This second patch is intended as a bass wave-sequence, and as such, I would expect only a single note to be held. You will hear that this patch doesn't work too well if you try to play chords.

There are two elements to this sound:

- the wave-sequence which is quite bright, and

- a filtered sawtooth which gives some weight and thickness to the sound.

To create the wave-sequence, I selected the Cosine Octaves wave from the Wavetable > Generated group of waves and set the Shape to be modulated by LFO 1 in which I selected the Step Generator. I tempo sync'd the Rate control to 1/16ths and set the steps to varying levels to create the wave-sequence.

Also within the Step Generator, I selected the black buttons at the top of the Steps. With these selected, the filter envelope and the volume envelope will both retrigger on each step. In the Amp EG, I set the Decay time to 0.36 seconds and the Sustain level to zero. This creates quite a staccato sound on each step.

The filter took slightly more setting up. First off, I selected the Dual One configuration and sent oscillator one's output to the left channel (so that 100% of its output goes through the filter). For filter one I selected the low-pass Ladder filter and set the Cut-Off to around 37 Hz and the Resonance to 29%. I then adjusted the filter envelope:

- Attack time was kept at zero

- Decay time was set to 0.25 seconds

- Sustain level was set to 30%

- Release time was kept at its default, and

- F1 slider (the amount by which the envelope modulates the filter) was set to 75 semitones.

Remembering that the filter envelope is retriggered on each Step, the combination of the different waveforms and the retriggered filter gives a very animated, quite insistent sound.

At this point I turned my attention to the second oscillator into which I loaded the Classic waveform. The first thing I did was to direct this oscillator to filter two so it is treated separately to oscillator one. I then tweaked the Shape, Width, and Sub Width sliders in order to thicken up the sound. Once the sound was thick enough, I set filter two's Cut-Off to 25 Hz and its Resonance to 27%, and pushed the F2 slider in the Filter EG to 83 semitones. Using the filter in this way helps to shape the tone, and gives a solid foundation to this sound.

With the foundation in place, I returned my attention to the wave-sequence and to add some more tonal movement, I set LFO 2 to control the cut-off frequency of filter one. I set LFO 2 with a very slow rate and it can modulate the filter cut-off over a wide range giving a constant shifting tone.

Finally to complete the sound, I pushed the Feedback slider up to 27% and in the Waveshaper selected the Sinus option and pushed the slider up to 9 dB. The combination of these two added a nice amount of gritty brightness, getting me to a point when the sound felt "finished".

One thing you might want to try is to set the level of oscillator one to zero and then modulate this level with an envelope in the LFO section. You could either use the Delay time—perhaps set to one bar or two bars—so that the wave-sequence suddenly bursts in, or you could use the Attack time so that the sequence gently fades in. Try it and see which effect you prefer.

03 Sequenced Decay and Echoes

This third wave-sequenced patch in Surge is intended to give the effect of the natural tonal decay of a note while using wave-sequence techniques. You will see that this patch does not use a filter.

For this patch, I have selected the Guitar AC 1 waveform in the Sampled group of waves. As you play through these waves, you get the sound of an acoustic guitar note decaying. I could have swept through these waves with an LFO or an envelope, however, my preference was to use the Step Generator. One of the reasons I decided against this was that I wanted to create an echo-type feel with each echo being a further step in the note's decay.

To achieve this effect, I set the Rate in the Step LFO to 1/8ths and drew a Step pattern so that as the Step Generator progresses, later waves in the sequence are selected. I also selected the retrigger option so that the volume envelope is retriggered on each step. The Amp EG was then set as follows:

- Attack time was set to zero seconds

- Decay time was set to 0.25 seconds

- Sustain level was set to zero, and

- Release time was left at its default.

The combination of the Step Generator and the Amp EG gives a staccato echo-like decay to the sound. I then reinforced that sound by adding some timed echoes in the first FX send slot. The time for the left echoes was set to 4/16ths and the right echoes were set to 4/4ths. I then tweaked the Feedback/EQ settings until the echo sat nicely within the sound.

More Advanced Wave-Sequencing

There are several limitations in using Rhino and Surge for wave-sequencing. These can all be overcome by using Wusikstation which, having two wave-sequence layers each with 32 wave slots and 12 controller lanes (which can also act as modulation sources), was designed for wave-sequencing.

However, this flexibility with Wusikstation illustrates one of the key challenges with wave-sequencing: the endless permutations. Wusikstation has several hundred waves available (indeed, maybe several thousand by now). The permutations of these several hundred waves are endless. Once the waves have been selected, they must be sequenced—again the permutations can be quite daunting.

However, once a series of waves has been chosen, the process for selecting the waves is quite straightforward in Wusikstation (even if the permutations are numerous). If nothing else it is much easier to choose a wave in a Wusikstation wave-sequence layer than it is to draw a wave-sequencing envelope in Rhino.

The following example patches are all created in Wusikstation. The patches have all been created using the Famous Keys waves that are included with Wusikstation—the full waveset may not come with the demo and so if you are using the demo of Wusikstation you may find that these patches do not work as intended.

01 Waveseq One

This first patch sequences 16 waves together. The waves are DW8001 to DW8016 which can be found in the Famous Keys Volume Three soundset.

The sequence has been set to step forward on each quarter note and plays wave 1 to 16 in order.

For me, the main purpose of this patch was to audition the waves and to listen to how they sound when put into a wave-sequence.

02 Waveseq Two

Waveseq Two is very similar to 01 Waveseq One—the waves are the same and the order in which they are played is the same. There are two key differences:

- First, the waves cross-fade (in the second controller lane xFade has been set to the maximum for each step), so instead of there being a step in tone with each new wave as in the first patch, with this sequence there is a constant shift in tone.

- Second, the time for each step has been increased to be equivalent to a half note. Without this increase, the fade between waves felt too abrupt.

As with 01 Waveseq One, the purpose of this sequence is to listen to how waves sound when put into a wave-sequence.

03 Waveseq Sustained Pad

Now that the waves have been auditioned, I want to construct a pad sound. I used some of the waves from the previous two patches to create this pad, but so that I didn't have to reload the slots, I have simply copied the previous two patches and only chose the waves I wanted.

The step length is set to 1/8 and as you will see, each step has its time set individually (in a separate controller lane) so that this does not become a regularly changing pattern. Also, as this is a pad, I didn't want there to be any stepping so I ensured xFade was set to the maximum for each step.

The first wave is DW8010 (which is in slot 10). This has a sort of buzzy electric piano type tone—I used this to give the wave some brightness in its attack. As you will see later on, I have used the filter to slow the attack of the brightness and so the piano like envelope will be masked. I have set the time for this first step to three, so it will last 3/8.

To follow this wave I chose DW8004—there was no real logic here, it just sounded right to me and tonally it fitted well after the previous wave. The time for this wave was set at two.

The next wave is DW8007—as with all of these waves it was chosen for its tone. The time allocated to this wave is four.

The fourth wave is DW8013 and the time allocated is 10.

The final wave is DW8016. As the intention is that this wave sounds for the whole of the sustain portion of the note, the wave has therefore been included in several slots each with the maximum time selected—this way the sound will sustain rather than loop back to the beginning of the sequence.

Having set the waves, I added a filter to give the player a bit more control over the sound. The filter does two things:

- it allows the player to control the tone of the pad, particularly during the attack phase, and

- it also acts as a quasi volume control to quieten the pad during its sustain phase—you will have noticed that there is no conventional volume envelope setting for this sequence.

Mod Env One is set up to control the filter—the modulation of the filter is controlled through the modulation matrix. The Mod Envelope achieves three things:

- first it makes the envelope's control of the filter fully velocity sensitive, so at higher velocity levels the filter opens more

- second, the Mod Envelope opens the filter slowly, having the effect of slowing the attack of the wave, and

- third the envelope closes slowly (the decay time is set to the longest possible time and the sustain level is set to zero) using the filter as a volume control. However, the filter does not fully close as the minimum amount of modulation in the matrix is set to 25.

Lastly, to add a bit of movement to the pad, I have added an LFO to give some vibrato to the sustain portion of the pad. This is done with Mod LFO One which is controlled through the modulation matrix by Mod Env One set with a long attack time so that the effect of the vibrato can be faded in.

04 Waveseq Rhythm Pad

Now that we have created a sustaining pad with a shifting tone, it might be interesting to create a rhythmic pad as a contrast.

Rhythmic wave-sequences can be quite effective for single notes. For chords (as would be typical with a rhythmic pad) the effect can be overpowering, so to get a rhythmic wave-sequence pad to work it has to be restrained. This patch has been designed to be played with chords in the mid-range of the keyboard: you will find it doesn't sound great when played outside of this range.

You will also find that this patch leaves lots of room for other elements in the production. However, it does not work well if you have a rhythm that simply plays eighth beats. To use this sequence in a track you will need to be a bit more creative in your rhythm programming.

The waves for this pad are again the same as those in 01 Waveseq One and as before I have only selected those that I want. However, unlike with 03 Waveseq Sustained Pad, the waves in this patch have been selected on the basis that they give a rhythmic transition from the previous wave. And of course, to keep the rhythmic effect, xFade is not used.

The other key feature of this sequence is the two silences. These are an integral part of the rhythm which you will have noticed is not a regular beat.

I won't take this pad apart and describe the individual waves in detail—those of you who have purchased the patches can take a look.

05 Waveseq Bass

We've built a rhythmic pad, so now it is time to build a rhythmic bass patch. This patch, 05 Waveseq Bass, is built around the waveforms chosen from the same range of waveforms selected for 01 Waveseq One. However, for this patch, I am also going to thicken up the sound by adding two conventional oscillators—I will explain this later.

The pattern for this wave-sequence is a regular sixteenths rhythm. The waves that I have chosen were those that gave an element of drive, but in the lower regions of the keyboard.

To my ear this choice of waves gives a good rhythm, but the sound lacks real weight. The sequence is great in a certain range of the keyboard—above the range and the patch sounds a bit thin and metallic while below the range the patch sounds to muddy and ponderous. For me the way to add weight is to bring in another tone.

To add this extra weight, in oscillators one and two I called up the saw waveform—both oscillators have had their pitch dropped by an octave and are then slightly detuned by reference to each other and panned slightly to the left and slightly to the right.

However, these two saw waves dominate the sound too much and cut in too quickly. To counter this I closed down a low-pass filter for each of the oscillators. These filters are then controlled by Mod Env One which takes some of the bite off the filters as the envelope is set to slowly open. In the modulation matrix, the amount that the filters are opened is slightly different for each filter giving a slightly different tone for each layer. However, the amount that the filter opens is not that much—this dull tone ensures the saw oscillators add weight without becoming a dominant feature.

The last tweak I have added is to drop the pitch of oscillator one when the key is first struck and allow the pitch to rise as the note fades in. This pitch change is controlled by Mod Env Two. The effect it gives is like an engine revving. When the oscillator reaches the right pitch, the sound of the oscillators acting together gets fatter and warmer, making an excellent backing for the rhythmic wave-sequence which sounds from the moment the key is pressed. You will find that this is the sort of patch where notes need to be held and any changes between notes should be carefully planned.

06 Split Waveseq

The last wave-sequence uses a different set of waves. This patch is also split.

- The lower half of the keyboard controls the wave-sequence layer. Note that this is a monophonic patch (so don't try chords).

- The upper half of the keyboard introduces a stab element to the patch.

There is also an overlap: the top octave of the wave-sequence layer and the bottom octave of the stab overlap.

Layer one and layer two are quite simple, both being based on Super Saw waves (from the Famous Keys Volume 2 bank), one slightly detuned from the other. These layers are intended as a stab patch which can be played while the wave-sequencing bass notes are held.

Turning to the wave-sequence layer, this is based on three waves: Super Saw and Analog201 (from the Famous Keys Volume 3 bank) and Spectrum1 (from the Famous Keys Volume 1 bank) which gives the metallic sounds.

The first lane of the wave-sequencer controls the soundset (the wave) that is played—for this patch the waves play in the following order: Super Saw, Analog201, Spectrum1, Super Saw, Analog201, Super Saw, Spectrum1, Super Saw.

The second lane controls the xFade—slots four and seven are set to the maximum, all other slots are set to zero. Slots four and seven gently transition to their new waves. For the other slots, the transition is immediate (and therefore, in the context of this patch, sounds more rhythmic).

In lanes three and four, the controller act as modulation sources. In this case the sources are routed (in the modulation matrix) to the filter Cut-Off and Resonance respectively.

FM Wave-Sequencing

If you're really up for a programming challenge, then Rhino allows you to achieve something more subtle than wave-sequencing: FM wave-sequencing!! Instead of creating regular FM sounds you can sequence the modulators. This has a lot of advantages:

- You can smoothly transition from one sound to another.

- You can achieve the wave-sequencing effect with sine waves alone—you don't need to use other waveforms (although that being said, if you're into experimentation, you can try FM wave-sequencing with different waves).

- You are not limited to a set number of tones where if you use Rhino conventionally, you can wave-sequence with only six waves.

With FM wave-sequencing in Rhino you have a maximum of five modulators. However, you have control of the levels, so immediately you have more than five basic tones. Secondly, you can arrange the modulators in parallel which gives you a much wider range of tone and you can modulate the modulators—I wouldn't say you have an infinite number of tones available, however, you certainly have more possibilities than you could work through in a couple of days.

Also, with the constantly variable envelopes, you have much more flexibility in how your tone changes—for instance if you have two modulators acting at once, one could stop immediately and the other could gently fade: this sort of tonal change cannot be replicated with conventional wave-sequencing techniques. You can, of course, perform FM wave-sequencing with Wusikstation, however, its FM implementation is not as readily usable as that under Rhino.

If you want to take the concept further and are confident about addressing the tuning challenges with FM, then you can even create FM wave-sequencing arpeggios.

As you would expect, you can also create these sounds in Surge using its FM capabilities coupled with its envelopes and the Step Generator.

A few examples may help to explain the technique.

FM Wave-Sequencing Patches

If you haven't done so already, I suggest you get hold of the patches before you read this section. The patches all use quite detailed envelopes to achieve their effect and it may be quite tiresome to manually recreate this programming.

The following patches are all variations on a theme. When building these patches I was aiming to create a driving bass rhythm that needs to work in the context of a fairly sparse track: a simple drum pattern, this bass part, and a lead line.

With a track this sparse, the bass needs to fill a big space. This means that the bass part will need to be (comparatively) busy and contain a fair amount of mid-frequency information. However, if the bass is to take up such a big space in the sonic spectrum, then is doesn't need really low elements otherwise it may smother some of the drums (the kick drum in particular).

Setting up the Patches

The patches are built around one carrier and four parallel modulators which have individual envelopes applied to give the wave-sequencing rhythm. For the first patch, the fourth parallel modulator is not included. No FX and no filters were used in the building of these patches.

- Operator one is set as the carrier. There is no adjustment to the Coarse Pitch and the envelope is a simple on/off setting. If you mute the modulators, you will hear a plain sine wave.

- Operator two is set at the same pitch as the carrier—this gives the tone considerable weight (although of itself, this tone is not that interesting). The fun stuff happens when the envelope is introduced into the proceedings—this is described below.

- Operator three is pitched one octave below the carrier (the Coarse Pitch is set to -12). The effect of this modulator is to add depth and bass to the tone of the note. Other modulators pitched below the carrier could be used to give a fuller bass sound, however, they may also lead to tuning difficulties, so for this patch I have chosen not to include any difficult elements.

- Operator four is pitched 19 semitones above the carrier (so the C:M ratio is 1:3). The purpose of this modulator is to bring some bite to the sequence.

- Operator five is pitched 24 semitones above the carrier (giving a C:M ratio of 1:4). This adds a touch of brightness (especially in the later patches).

FM Wave Seq 1

We've seen the main elements for the sequence. This is how the sequence is put together.

Operators two and three are set to fully modulate the carrier, operator one. The amount by which operator four modulates operator one is set to 75. For each modulator the Basic 16 envelope was selected and then tweaked. Mute the effect of operators three and four on the carrier before continuing to look at this patch.

The effect of operator two is to give the patch tone and drive: it doesn't add much bass to the sound. The sixteenth rhythm plays fairly consistently, although on steps two and six the amplitude is reduced to 50% and on steps four and five the envelope is set to zero (so only the tone of the carrier is heard). The pattern played by operator two is quite repetitive—I felt the patch needed this to keep its drive. However, you will hear that some of the repetition is counteracted by the effect of operator four.

Where operator two is busy, operator three creates a pattern which is quite sparse, only playing on four individual steps. While only playing on four beats, there is another change to the rhythm with this operator in that some of the steps sound for the whole length of the beat rather than being simply an impact.

The sequence created by operator three may be sparse, but when operators two and three work together it creates a good rhythm, but the end of the pattern is still a bit repetitive. The tone of the pattern also lacks some variation. To address these shortcomings, operator four is called in.

Operator four has five steps and some of these last for the beat (rather than just for the impact). Operator four also acts to break some of the monotony of operator two without detracting from the rhythm established by that operator by putting all of its beats (except the first) towards the end of its sequence.

The final tweak I made to this patch was to assign a user slider. I'm not a great fan of this feature if it is just linked to one control. However, it can be really useful when several parameters are controlled simultaneously by the slider (especially if the slider is then controlled by some external hardware with the MIDI learn function). In this case, the slider marked Intensity controls the amount of modulation by operators two, three and four.

You will see that each operator has a different range: operator two's modulation goes from 45 to 100, operator three's from 65 to 100 and operator four's from 15 to 75. To my ear, these ranges gave the broadest spectrum of usable tones. A practical use for this feature might be at the start of a track—if you set the intensity slider to be controlled by a track envelope, then at the start of the sequence you can gradually increase the intensity, perhaps over two bars when the drums could then join the track.

FM Wave Seq 2

This next sequence builds on FM Wave Seq 1. The key changes are:

- operator five has been added as a fourth parallel modulator
- all modulation levels have been set to 100, and
- the intensity slider has been dropped.

In this patch, operator five adds a touch of emphasis to some of the beats. Its effect is quite subtle having more of an effect on the tone and texture of the patch rather than the rhythm. If you listen closely you will be able to hear that on one beat the operator fades in (very quickly)—again the effect is very subtle but it is almost like applying reverse reverb to one beat.

FM Wave Seq 3

Tonally FM Wave Seq 3 is much brighter than FM Wave Seq 2. There are two reasons for this brightness:

- In operator five, the sine wave has been replaced with a triangle wave—this gives the effect of operator five more sizzle.

- Operator two (which modulates operator one) is now modulated by operators four and five. This changes the waveform in operator two that is modulating operator one and so brings more tonal variations to the pattern. In effect we're wave-sequencing twice in a three operator cascade, which has parallel modulators at its source. See Figure 8.1 for an illustration of the signal flow.

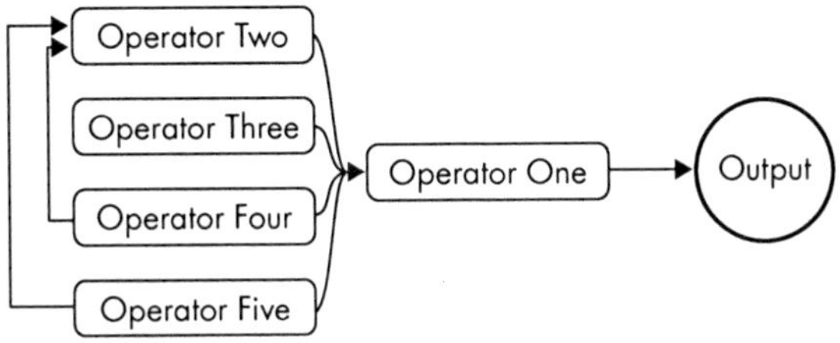

Figure 8.1 The routing adopted in FM Wave Seq 3

I have assigned some sliders to allow for some tonal variations and so that you can easily hear the effect of the extra modulations.

- The slider labeled Raw 2 & 5 allows you to directly feed the raw outputs of operators two and five. I was hard pressed to hear any difference in the tone by introducing these raw waves—however, you can hear the effect when some of the other modulations are reduced in level.

- Operator 2 Mod allows you to reduce the amount by which operator two is modulated by operators four and five. When this is set to zero, you can make a direct comparison with FM Wave Seq 2 to hear the effect of changing operator five's wave to a triangle.

- 3, 4 & 5 Mod 1 controls the amount by which operators three, four and five modulate operator one. If you set this to zero it is easier to hear the effect of the changes controlled by the Operator 2 Mod slider.

FM Wave Seq 4

This is the final patch in the series. It has the most tonal movement and is the brightest. However, that does not necessarily mean it would be right for your project. Within the track for which this patch is intended, the brightness works.

As you would expect, FM Wave Seq 4 is a development from FM Wave Seq 3. The main changes are:

- The envelope for operator two (which gives much of the drive to the patch) has been restructured. The rhythm has been kept but with a few tweaks added here and there (especially to some of the curves). My method for deciding what I wanted to change was to listen to the sequence looping and then make a slight change where the patch felt repetitive. As always, you may not agree with the choices I made.

- Three new modulations have been added:
  - feedback has been added to operator two (which modulates the carrier)
  - a touch of feedback has also been added to operator three (which also modulates the carrier), and
  - operator five modulates operator three.

  The effect of these modulations—particularly on operator two—is to give a greater range of tone shift as the sequence plays.

- The three modulation tweaks can be controlled by a new slider labeled New Mods.

- There is also another slider, labeled Raw 2, 3 & 5, which controls the raw output of operators two, three and five. As you increase this, the sound gains a slight element of presence: my preference is to set the slider to zero and just rely on the carrier to provide the whole sound.

Once you have listened to the patch, if you have the feeling that this could have been inspired by a well known keyboard player of East European origins who scored a popular 1980s US TV cop show, you would be correct.

FM No Seq 4

To give you an idea of the sound of this patch without the wave-sequencing, FM No Seq 4 takes the previous patch and replaces the rhythmic envelopes with piano-type envelopes. You will see that a range of sliders have also been assigned to control the tone.

FM Arpeggio Patches

With a simple FM stack, when the carrier : modulator ratio is higher than 1:2 (for instance 2:5, 1:3 etc) the pitch of the FM patch is determined by the carrier. When the C:M ratio is less than 1:2 (for instance 1:1.5, 3:1 etc) the pitch is more influenced by the modulator. This gives us the opportunity to create melodic figures (such as arpeggios) in FM patches and at the same time have the tone shifts offered by FM wave-sequencing. However, please remember that the perception of pitch can be quite subjective, especially when an FM wave contains many frequencies which may not be harmonically related.

It is probably easiest to illustrate the principle with a simple patch.

FM Wave Seq Octaves

This patch is based on a parallel modulator arrangement.

- Operator two is the first modulator. It is pitched 12 semitones below the pitch of the carrier, operator one—its effect is to darken the tone and to drop the pitch of the resulting note by an octave. The Basic 8 envelope has been used and this has been tweaked so that the operator kicks on the beats.

- Operator three is the second modulator. It is pitched at the same frequency as the carrier and so when this note sounds it will be an octave above the note produced when operator two modulates the carrier. Again the Basic 8 envelope has been called up, but this time the peaks have been left on the "and" beats (that is peaks one, three, five and seven have been deleted).

Hit any note and you will get that classic root/octave bass line. Find a 1970s disco track and you're made for life.

FM Wave Seq Arpeggio

Its time to get a bit more adventurous: FM Wave Seq Arpeggio creates a simple arpeggio using the same principles established in FM Wave Seq Octaves.

Operator one is set as the carrier. Operators two to five act as modulators and create the arpeggio tuning as follows:

- operator two is tuned two octaves and five semitones below the carrier (giving a C:M ratio of 7:1) and acts as the root of the arpeggio

- operator three is tuned a major third above the root modulator (giving a C:M ratio of roughly 5:1)

- operator four is tuned a perfect fifth above the root modulator (giving a C:M ratio of 4:1), and

- operator five is tuned an octave above the root modulator (an octave and five semitones below the pitch of the carrier, giving a C:M ratio of 3:1).

Operator six is tuned to the same pitch as the carrier and only modulates it to give a slightly brighter tone to the arpeggio, in this case the modulation amount is set at 30.

Once the modulators had been set up, and the matrix was adjusted so that each of the four main modulators modulates the carrier by 100, the sequence was set up with the envelopes in each operator so that the modulators play in sequence to create the arpeggio.

Those of you who have the patches can hear the results. A very basic, quite uninteresting arpeggio is played. You can make the tone somewhat more interesting by playing with the two sliders:

- Weight increases the modulation of each of the four main modulators by operator six. It also feeds the raw output of the four main modulators to the output. You will see that this slider operates to increase the modulation only slightly (the maximum increase is 25) but the increase for the raw outputs is much more significant (70).

■ Brightness increases the modulation of the carrier (operator one) by operator six, and introduces some feedback into the carrier. As you would expect from the label on the slider, the effect of these two additional modulations is to make the sound brighter, however, I find the brightness works better when the weight slider is also increased—my preference is to set the weight slider to 40 and the brightness slider to 30.

A Few Thoughts about Wave-Sequencing

Wave-sequencing is somewhere between painting and sculpture—you keep adding shades and tones while chipping other pieces away. Frequently you need to stand back and look at what you're doing to make sure the overall effect is right.

It can be one of those instant gratification techniques. However, the technique does have its limitations. For instance, you can usually only use one rhythmic wave-sequence in an arrangement at any one time or else the track will just become too cluttered and confusing.

I like the technique because it can create rhythmic patterns which can only be created by using this technique: you cannot program a sequencer to create these sounds. I find the constant shifting/cross-fading techniques can generate interesting textures; however, often trying to sequence together several waves to create an interesting and cohesive texture just gets the same effect that can be achieved with conventional FM or subtractive synthesis (or a combination).

Wave-sequencing can take a while to achieve satisfactory results. However, I would encourage you to think of wave-sequences as being an integral part of your track's arrangement. If you follow this logic, then you should apply the same care to creating the sequence as you would when programming another other part within your track.

I think the FM wave-sequences add another option for creating interesting and compelling rhythm tracks. However, I am less convinced about whether the FM wave-sequenced arpeggios have much of a future. The programming of the arpeggios is long-winded and tedious and as the example above demonstrates, the tunings can be quite imprecise and then the tonal variations are limited as the tuning of the arpeggio can be affected by the modulation. At the end of the day, I do wonder why you would do this in Rhino when it already has a step sequencer and pitch envelopes.

Chapter 9
Additive Synthesis

I have spent a lot of time looking at subtractive synthesis where you take a sound source which is rich in harmonic content and then reduce and shape those harmonics with a filter.

Additive synthesis takes the opposite approach—it takes a bunch of sine waves at different frequencies and puts them together to create a new sound. The sine waves (or "partials" as they are called—the term partials and harmonics are pretty much interchangeable) are all multiples of the fundamental frequency.

Additive synthesis requires patience to construct these new waves, but the results can be excellent. There are many advantages to this approach:

■ Each partial can be individually controlled.

■ The amplitude of each partial can be controlled over time (depending on the architecture of the individual synthesizer).

■ Individual partials can be detuned (to introduce more metallic elements into a sound) and the phase of each partial can be controlled (again, both of these are subject to any limitations imposed by a synthesizer's architecture).

■ True morphing can be achieved.

None of these features can be created by a conventional subtractive synthesizer or sampler. In addition, some synthesizers (an example being Cameleon 5000) allow you to re-synthesize sounds.

Morphing and Cross-Fading

There is an important difference between morphing and cross-fading.

With cross-fading there are two sound sources—the effect is achieved by the level of one source being reduced while the level of the other source is increased. It is therefore possible to hear two distinct sounds during the process.

With morphing, there is one sound source—initially it has the characteristics of the first source and at the end it has the characteristics of the second. The composition of the partials making up the wave changes during the morphing transition. So when the sound is halfway through its morph, there is a wholly new sound rather than there being the two existing sounds as would be the case if the sounds were simply cross-faded.

In short, morphing allows for completely new sounds to be created. These sounds cannot be created by other means.

Warning!! Mathematics Ahead

This isn't a book about mathematics and if you want you can skip these next few paragraphs that give some of the theory behind the creation of waveforms with additive techniques.

I mentioned earlier that waves are made up from a combination of sine waves—different frequencies and different proportions of waves will give different results (or as we prefer to call them, different sounds). Without getting into too much detail of the mathematics, I will now set out how the three most common waves can be constructed using additive principles.

For those of you who are really keen on mathematics, you can work these formulae through and prove that the addition of these sine waves creates the resulting waves. For the less mathematically inclined (this author included), I have included graphs to show the amplitude of each harmonic within a wave—compare these graphs to some of the sine wave compositions in Rhino and Cameleon 5000 when we start creating waves from first principles.

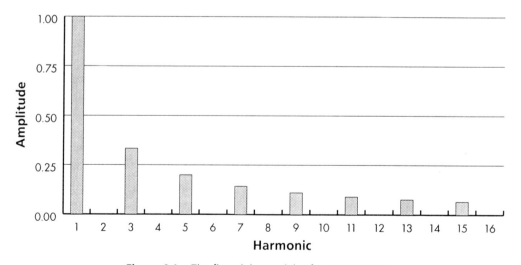

Figure 9.1   The first eight partials of a square wave.

Additive Square Wave

A square wave is made up of only odd-numbered harmonics (sine waves) with decreasing amplitudes in the ratio 1/n (where n is the number of that harmonic). The harmonics are therefore:

- the first harmonic (the fundamental) which has its full amplitude

- the third harmonic which has an amplitude of one-third of the maximum

- the fifth harmonic which has an amplitude of one-fifth

and so on.

Figure 9.1 plots the amplitude of each harmonic against the harmonic's position. However, you will note that this graph only shows the first sixteen harmonic positions where there should be an infinite number of harmonics—later on we will look at the relationship between the number of partials and the tone.

Additive Sawtooth Wave

A sawtooth wave has both odd-numbered and even-numbered harmonics with amplitudes decreasing in the ratio 1/n. The harmonics making up the sawtooth wave are therefore:

- the first harmonic at its full amplitude

- the second harmonic at half its amplitude

- the third harmonic at one-third of its full value

- the fourth harmonic at one-quarter of its full value

and so on. Take a look at Figure 9.2.

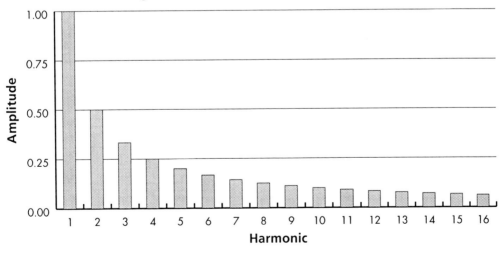

Figure 9.2 The first sixteen partials of a sawtooth wave.

Additive Triangle Wave

A triangle wave has odd harmonics with decreasing amplitudes in the ratio $1/n^2$. Therefore a triangle wave has the following harmonics:

- the first harmonic at its full amplitude

- the third harmonic at one-ninth of its full value

- the fifth harmonic at one-twenty-fifth of its full value

- the seventh harmonic at one-forty-ninth of its full value

and so on, as Figure 9.3 shows.

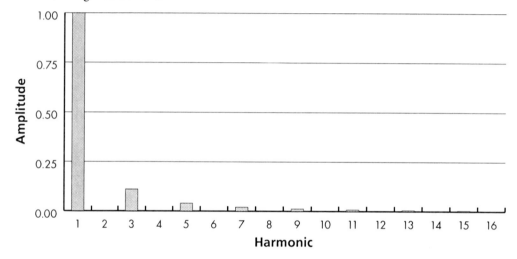

Figure 9.3   The first eight partials of a triangle wave.

As you will read, one of the themes of this chapter is to program with your ears, not your eyes. If you look at the triangle wave graph and the square wave graph, you will see some similarities—they both have odd harmonics declining in amplitude. Now listen to a square wave and a triangle wave—the difference is far more marked than the variations in the graphs may suggest.

If you want to delve further into the science of waveforms, then go back to the internet and search for "fourier synthesis".

Difficulties with Additive Synthesis

Like FM, additive synthesis has a reputation for being "difficult". "Difficult" may be an understatement—additive can be potentially hugely complex. Take sound creation in Cameleon 5000 as an example:

- each wave can have 64 partials—each partial has its own level and its own tuning

- at each point in the envelope, the partial levels and their tunings can be repositioned, and

- for each wave, there is an accompanying noise source with many options for control.

To then multiply the confusion there are four oscillators which you can morph between, remembering that the sound for each oscillator is constantly shifting.

If you adjust only 10 partials at three envelope breakpoints for the four waves, and add some morphing then you will have made nearly 150 adjustments.

Before you can get to the complexity and the endless permutations, there is a far more straightforward challenge—creating an interesting sound. If you can't create an interesting sound, then the issues of how you control the sound over time and how you balance several sound sources are irrelevant.

Unfortunately there is no magic formula for creating an interesting sound from first principles. You really need to sit down and listen and build up some experience in creating these sounds. However, you can always import waves (with Rhino you can import single cycle waves, with Cameleon 5000 you can import multi-samples).

If you're looking for a real challenge, then try creating some FM patches with additive waves: use an additive wave as a modulator and listen to how the changes to the wave affect tone.

Example Additive Patches

To help understand what additive synthesis can do and the tones it can create, it may be helpful to create some sounds using additive synthesis techniques. This first group of sounds will be built using Rhino. While Rhino's additive synthesis features are nowhere near as controllable as those under in Cameleon 5000, they are still very powerful and very usable making it the ideal tool to demonstrate some of the basics of additive synthesis.

The purpose of these patches is not to create great examples of sounds that can be made with additive synthesis. Instead, these sound are here to illustrate some of the fundamentals around creating additive waves.

These patches are all created with additive waves and envelopes—no filters or FX are used here.

Additive Build

This first patch, Additive Build, uses wave-sequencing techniques to develop the tone of the note over time. The patch will sound rather like an arpeggio except that all of the earlier notes are held as higher harmonics are introduced to change the tone.

When you strike the note, you will hear the first oscillator—in this oscillator an additive wave has been created by choosing the first harmonic (in other words the fundamental) only. You will next hear the second harmonic added—the level of this second oscillator has been set in the matrix at 55%.

The third harmonic (three times the frequency of the fundamental—19 semitones above the fundamental) is the third wave to be added in the series and comes in at 40% of the level of the fundamental. The fourth wave is the fourth harmonic (four times the frequency of the fundamental or 24 semitones above the fundamental) and comes in at 35% of the level of the fundamental. The fifth and sixth harmonics then follow at 20% and 15% of the level of the fundamental.

So what have we done here? We have created a wave by layering harmonics—you can hear that as each harmonic is added the tone of the wave becomes brighter. If you want to hear the tone of the note once the harmonics have been added but without the sequential build up, you can listen to the next two patches:

- Additive Build [Tone] is a copy of Additive Build, the only difference is that the envelopes are all the same and now all of the waves all play together at the same time (with the same mix levels as before)

- Additive Build [Wave] recreates the tone of Additive Build [Tone] by selecting the individual partials in the additive wave generator. You can hear the sound is the same as Additive Build [Tone], but you will notice that the main volume fader has been increased to balance the single oscillator against the six oscillators in the previous patch. The particular advantage of this approach (compared with Additive Build [Tone]) is that only one wave is used—this uses less CPU and leaves the other five waves free to create tones (or to layer/detune the same wave in another oscillator).

So as you can hear, we have taken six harmonics and created something that sounds quite similar to a sawtooth wave. Who said additive was difficult?

You will notice if you look at the matrix in Additive Build that each of the waves is progressively quieter. However, if you listen to the waves as they enter in sequence there is little perceptible difference in volume of each of the waves—certainly oscillator six does not sound as if its volume is 15% of oscillator one's volume.

This highlights two issues:

- first, the disproportionate effect the higher harmonics can have, and

- second, it is really difficult to adjust those high harmonics—you only want to move them one or two clicks and this is difficult with a mouse and the GUI doesn't really give enough immediate feedback.

One other thing you may notice—if you listen to Additive Build (when the six oscillators are heard) and Additive Build [Tone] it may not be immediately apparent that they are the same sound. With any shifting sound source it is always difficult to isolate one element of the sound for comparison.

Additive Square and Better Drawn Add Sq

Here are two examples to illustrate how you can spend a long time trying to make an additive wave look right when you should be listening to how it sounds.

First listen to Additive Square—this wave was created by importing a square wave. As you would expect, it sounds like a square wave. Now take a look at Figure 9.4 and notice the harmonic com-

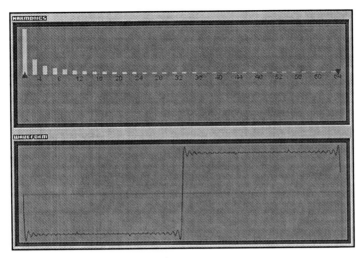

Figure 9.4   An additive square wave.

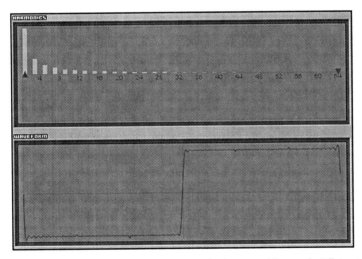

Figure 9.5   An additive square wave—note the improved image, but listen to the degraded sound.

position of the waveform. You will see that the wave includes all of the odd harmonics (with the amplitudes described earlier).

Now take a look at Figure 9.5. You will see that this square wave has a different harmonic composition—in particular, no partial above the 43rd partial is included. However, you will also see that the image of the square wave (below the partials) is far smoother and much more closely resembles a square wave.

If you listen carefully you can hear that this new wave has a different tone. I am not sure that the effort need to redraw the wave justifies any tonal changes—the change here is probably to the detriment of the sound.

169

While I can hear the difference when the waves are played individually, I'm not sure that there would be much difference in a patch, let alone in the context of a fully orchestrated track.

Square Type Sound

For this next patch, my intention was to create something from scratch that sounds (vaguely) like a square wave but only using three partials—I set myself the three partial limit largely because I can get quite frustrated when trying to manipulate more partials. As I was after a square wave sound, I chose the first, the third and the fifth partial—the first partial is set to the full amount, the third partial to about halfway, and the fifth partial to around 15%.

If you listen to this patch the sound does resemble a square wave. However, to my mind, this tone sounds more like a square wave through a filter, or rather more like a square wave *should* sound through a filter given that square waves don't react that well to filtering (in my opinion).

You will also remember that with Additive Seq it was harder to hear the final tone when the waveform was shifting (due to notes being added)—you may find a use for this truncated version of a square wave if you use it in passing rather than as a feature.

Reed Squ Type Sound

For this next patch, the mission I set myself was to take the previous three partial patch, Square Type Sound, and to make it more "reedy". To achieve this, I added an element of the second and fourth harmonics. The second harmonic was set to around 20% and the fourth harmonic to about 30%.

These two extra harmonics also lost some of the character of the square so I increased the amplitude of the third harmonic to around 75% and increased the amplitude of the fifth harmonic by a few percent. For both this patch and the previous patch, these levels were set by ear—there is no scientific logic called upon and the waveform display is of little interest.

As with the previous patch, this patch was not intended to replicate a specific sound, merely to create a tone color using a minimal number of partials. In this it has succeeded. Hopefully these two waves have illustrated that it is possible to create additive waves quite quickly and easily.

Phase I, Phase II, and Phase Shifting

Listen to Phase I and then listen to Phase II. Now tell me, can you *honestly* tell the difference between these two waves? I certainly couldn't.

Now look at Figure 9.6 and Figure 9.7. You should notice two things—the harmonics are exactly the same, but the phase of the harmonics is different (the phase of each harmonic is controlled by the lower half of the top pane). The result of tinkering with the phase is that the resulting waveforms look wholly different, but still sound the same.

As you might suspect, this is a patch which has been contrived to illustrate the point that you need to use your ears and not yours eyes. Also, to my mind it suggests that when you program additive waves it is fairly pointless to muck around with the phase of individual harmonics.

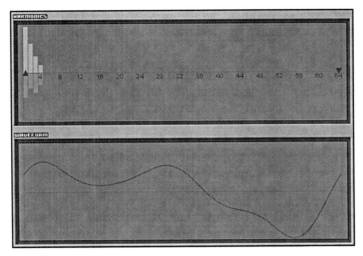

Figure 9.6 The additive wave used in the patch Phase I.

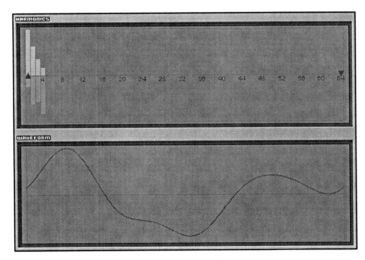

Figure 9.7 The additive wave used in the patch Phase II.

So if the waves sound the same, are they the same? No—listen to Phase Shifting. This has the Phase I wave loaded into oscillator one and the Phase II wave loaded into oscillator two, and the oscillators' envelopes have been set to cross-fade between the two waves. As you hold a note, you will hear tonal shifts. If these waves were identical, there would be no shift in tone.

If you're feeling adventurous, you can set the waves to play together (just tweak the envelopes to be identical)—here you will hear a tone that is different from the tone you would get if the waves were the same: in particular, if you listen to the higher frequencies, when the waves play together the organ-like tone takes on somewhat more of a vintage character. The reason for the difference is the waves that are out of phase are partly cancelling each other out—this gives a different tone (this cancellation works very much like conventional oscillators that are not working in phase sync).

Bright Additive

We have created some additive waves, but with all of the control and tone shaping options we have only made dull imitations of conventional waves. Let's make a brighter wave that doesn't resemble any wave we have considered so far.

One downside with additive waves is that it is very easy to create organ type tones—another problem is that if you get away from an organ-like tone you can start to create something really wild. While a wild sound may be interesting, it is not necessarily useful, and neither is it controllable in a patch.

To keep this wave controllable, I only adjusted three partials—the fundamental, another to give a bit of body and the last (the highest) to give the bright tone. As a first step, I increased the volume of harmonic one (the fundamental) to 100.

To get the metallic brightness I was after, I increased harmonic 20 to 45%. I chose harmonic 20, because the harmonics below this gave more of an organ like tone when working in combination with the fundamental and the higher harmonics made the sound too piercing—harmonic 20 seemed to be the ideal compromise. I then increased it to a level that sounded right to my ear—in this case it was 45%. You may or may not agree with the choices I have made here.

Finally, to add a bit of weight to the sound I added a third harmonic. As with the choice of the brightness element, the choice of a harmonic to add weight was a compromise. Lower harmonics gave more of an organ quality while higher harmonics didn't add weight and could interfere with the sound created by harmonic 20. I did consider harmonic 12, but felt that gave too much of a reedy tone and so in the end I chose harmonic 8 (which is three octaves above the fundamental) and increased it to 60%.

As you can hear, with just three harmonics this wave has created a tone that has an FM-like quality. While the results were obtained to a certain extent by trial and error (at least in the early stages) creating a wave this way is far easier than creating a wave with FM.

If you are creating from first principles (in other words starting with an empty wave and adding harmonics) you will be able to clearly hear when any addition creates dissonance and does not add to the tone and remedy the situation immediately. The down side to this form of wave creation in Rhino is that all you are doing is creating a waveform—these waves are not controllable in the way that they are in Cameleon 5000.

That being said, the downside in Cameleon 5000 is that controlling the waves becomes a bit of a hit and miss affair—it is not immediately evident how each partial affects the tone when you are tweaking partials that occur later in a note's passage.

Additive Bell EP

Although it is interesting to create tone with the additive wave generator in Rhino, the results still haven't passed the "so what?" test. To be useful as a technique, the waves need to be capable of being used to create a sound which could be used in a musical context.

I used the waveform I created in Bright Additive to create an electric piano-type sound. For this patch, I loaded the waveform into oscillators one, two and three and then tweaked the additive waveform as necessary to get the tonal changes. I did not use a filter to create this sound but instead have relied on the three oscillators alone. Equally, no FX was used to create this sound.

The first thing I did was mess around with the waves:

- Oscillator one was left untouched—this is used to give the bell tone in the attack phase of the note.

- Oscillator two has harmonic 20 removed—this gives the note a more bright organ like quality.

- Oscillator three has harmonic 20 removed and harmonic 8 reduced to 15—while this does give an organ like tone, when used in conjunction with the bright attack sound from oscillator one, this gives the ideal sustain sound for our electric piano type patch.

With the waves set up, the **first** task was to take oscillator one and adjust its envelope to give the attack portion of the note and then to cross-fade that into the sustain portion of the note created by oscillator three. For both oscillator one and oscillator three I took one of the standard piano waves and tweaked it.

For oscillator one, the envelope falls rapidly to zero over a period of about 1.2 seconds. With a steep curve (in this case set to around 25), this gives the classic bell like tone to the attack of the note. By contrast, oscillator three has a similar curve, but its level falls to about half over a much shorter period of time—around 0.25 seconds. This envelope also has a sustain loop (which is not surprising as this wave is used for the sustain portion of the note).

The interaction of these two envelopes gives me a seamless cross-fade between the two sounds. The cross-fade has the smoothness of an FM tone change.

Both oscillators one and three also have an element of touch sensitivity which controls the volume of each oscillator.

These two oscillators alone give a serviceable, but simple, electric piano sound. However, I wanted a slightly more sophisticated tone and to give the player more control. This is where we call upon oscillator two, which has a tone that is brighter than oscillator three but duller than oscillator one.

A very simple envelope is used for oscillator two—it fades in and then remains at its sustain level. However, the velocity curve for this oscillator is set so that it is only heard at the higher volumes, at lower volumes the oscillator has no effect on the sound. Finally, I have added a bit of detuning to thicken the sound with a natural chorus effect.

With the second oscillator, the patch has a thicker tone with more of a sparkle at higher velocities.

You may be wondering why I created *another* electric piano sound, given that we can already achieve this sound using FM. There are several reasons:

- This method gives a different tone—it is thinner and brighter—which may suit a particular arrangement, especially one which doesn't want to be dominated by an FM electric piano sound.

- This sound is far less clichéd than an FM electric piano.

- The use of less oscillators means that less system resources are called upon—also, as I've only used three of the six oscillators, it would be possible to double this sound for a thicker tone or to explore other sonic possibilities.

- It is simpler to design this sound and for the musician, it is easier to understand what is going on—the tone can be easily changed: there are no complex FM cross-relationships that mean one tweak could wreck a whole patch.

To give a more realistic emulation of an electric piano it would have been possible to route the three oscillators through a filter and use a filter envelope as a volume control. However, this option didn't seem to add much to the sound of the patch, so I didn't do it.

More Complex Additive Patches

So far this chapter has looked at some of the things that additive synthesis can do. The examples have used Rhino as the sound source and so we have created static waves whose composition of partials does not change over time.

Cameleon 5000 gives significantly more control to the additive sound creation process. For instance, with Cameleon:

- Each element to be individually controlled over time, so for instance, individual partials can be controlled over time, meaning a single wave can get "duller" as it sustains without using a filter.

- Noise elements can be added to the sound for more realistic sound creation possibilities.

- Individual partials can be detuned.

- Touch sensitive (or other modulation controller) morphing between two tones can be readily achieved.

- Multi-sample resynthesis can be employed to (re)create more interesting instruments allowing their timbre to change over the whole keyboard.

The focus of this book is on ground-up synthesis, rather than sampling/resynthesis, so many of the great features of Cameleon 5000 are being ignored (but many of the features of the other synthesizers are being ignored too). In particular, this book does not look at creating sounds on the basis of presets or resynthesized samples. Instead, it focuses on creating sound from scratch by manipulating of the individual partials and also creating sounds by using the limited supply of stock waves.

There are three disadvantages to the approach taken by this book:

- First, it is difficult to get usable results.

- Second, the stock waves are limited (for instance, none includes any FM or metallic type tones).

- Third, there is a tendency to create organ type sounds and if you wanted an organ, you would have bought an organ.

So why are we doing this? Simple—if you understand and can control the waves in this situation, it becomes much easier to control the more interesting waves.

01 Saw to Square [Envelope]

One of the key features of Cameleon is its ability to morph between sounds. This first patch, 01 Saw to Square [Envelope] demonstrates that morphing. As you strike a key you will hear a saw-tooth wave (the wave I have used is the Saw-Bright wave which was chosen from the list of waves in the harmonics drop-down). If you continue to hold the key this sound will progressively morph into the sound of a square wave (the square wave I have chosen in the harmonic drop-down is Square-J). As the key remains held, the sound will morph back to a sawtooth wave and the loop will begin again.

This morphing is achieved by drawing an envelope in the morph timeline.

While this may not be the most interesting sound, listen to the morphing. Then fire up another synthesizer and cross-fade between a sawtooth wave and a square wave and note the difference. Listen to the smoother tonal shifts in Cameleon 5000—with a cross-fade you can hear the separate elements whereas with the morph you hear one far more cohesive tone (albeit, in this case, one that could do with a good bit of filtering).

02 Saw to Square [Velocity] and 03 Square to Saw [Velocity]

Instead of using an envelope to control the morphing, the tone shifts in these patches are controlled by velocity.

For 02 Saw to Square [Velocity], at the lower velocity levels you will hear a sawtooth wave. At the higher velocity levels you will hear a square wave. Between the extremes you will hear the morphed sound. Listen to the different nuances as you play the patch at different velocities.

The second patch, 03 Square to Saw [Velocity], follows the same logic as the previous patch, however, the morphing goes from the square wave at lower velocities to the sawtooth wave at higher velocities.

Instead of using a sawtooth and square wave, you could set up two variations on a sound—a bright sound and a dull sound—and then use velocity scaling to morph between these two extremes

175

creating a natural response without having the sonic intrusions that occur when you can hear a different velocity layer being selected in a sampler. As we are morphing, there are no steps to worry about.

04 Additive Electric Piano

You will remember that earlier in this chapter we built the Additive Bell EP patch with Rhino. This patch uses a similar simplicity to create a similar kind of sound.

The patch uses four breakpoints—at each breakpoint the composition of the harmonics is redefined. As the patch plays and a note passes from breakpoint to breakpoint, the harmonic composition of the note morphs between the tones set at each breakpoint. The result is a constantly and consistently changing tone.

As a first step, I created four breakpoints and dragged the envelope to vaguely follow the shape of a piano envelope—the four points were set:

- at the peak of the attack

- immediately after the attack

- about two seconds into the decay (in other words after the initial tone shaping of the attack had ceased to have effect), and

- after about seven seconds.

These points were chosen to broadly mimic the four main stages of the note in a natural instrument.

At the first point—which corresponds to the peak of the attack of the note—I drew some harmonics which broadly followed those that I used in the Additive Bell EP patch in Rhino. I used the same harmonics (first, eighth and twentieth) and set the levels to broadly the same level as they were in Rhino. If I were trying to create every nuance of a realistic instrument, I would also have added some noise in the attack phase.

The second breakpoint marks then end of the decay phase which comes immediately after the attack phase—in a conventional ADSR envelope this point would determine the decay time and set the sustain level. For the tone to reflect the characteristics of a real instrument, it should be quite bright. Compare this approach to what would happen if we were using a conventional subtractive synthesizer—at this point the filter would generally have been closed down to the sustain level. However, as we are using additive synthesis, we can adopt an appropriate volume envelope while ensuring that the tone does not change significantly.

To give a bright tone, I have again used the same three harmonics, but to take account of there not being a hammer impact at this phase of the note, I reduced the level of the twentieth harmonic. If you look at the harmonic content in Cameleon 5000 (by engaging Breakpoint and clicking on the breakpoint) you will see that all of the harmonics are lower—this is to be expected as this phase of the note is quieter. However, if you lift the level of this breakpoint to be equal to the attack breakpoint and then compare the respective harmonic compositions, you will see that this breakpoint does have less high frequency information.

The third breakpoint is intended to mimic the tone about two seconds into the note when the effect of the hammer action has passed. There are two differences between this and the second breakpoint. First, there is no twentieth harmonic and second the eighth harmonic has been reduced further. As before, since this point is quieter in volume than the second breakpoint, both of the remaining harmonics have had their amplitude reduced.

The effect of these first three breakpoints is to give a constantly shifting tone to mimic the attack phase of an electric piano. By gently shifting the harmonic composition of the note, and its volume, a natural effect can be achieved. After this I have only added one further breakpoint at about seven seconds. The waveform at this point only contains the fundamental and so allows for the note to morph over time and lose its brightness as it decays.

I quite like the attack of this patch—in a track it could work fairly well. However, there are some shortcomings here:

- There is no player control over the tone. A copy of this wave could be created and the harmonics tweaked at each breakpoint to give a duller sound—velocity scaling could then be used to morph between the two waves to give a far more naturally responsive sound.

- While the attack is good, the sustain portion of the note is quite dull. This could be rectified by adding more breakpoints and changing the tonal composition of the note. Also an element of wobble could be introduced with an LFO to give a bit more life to the sustain portion of the note.

You may think it is too much of a delicate approach to cut out or reduce the higher harmonics and you may also rightly think that it is quite time consuming. Perhaps a preferable method of operation is to use the filter to remove some of the main high frequency elements and to "thin" the sound by cutting individual harmonics over time. Alternatively, you may think that using a filter for an additive synthesizer is cheating!

05 Organ

I've pointed out at several points that one of the drawbacks (no pun intended) of additive synthesis is the inadvertent ease with which organ sounds can be created. The flip side is that additive is a fast way to create interesting organ tones as this patch will show.

I have created this patch with three breakpoints. The first at the attack stage, the second immediately after the decay (in other words at the start of the sustain phase), and the third at the end of the sustain phase. The waveform at the first breakpoint is created by combining the first four harmonics to give an organ-like tone.

At the second breakpoint (at the start of the sustain phase), the overall volume of the envelope is reduced, but a fifth harmonic is added to give a touch more brightness to the organ tone. The third break point (at the end of the sustain phase) is set at the same level as the second breakpoint (so there is no change is volume as the note sustains), however its tonal content is much closer to the first breakpoint, in other words it is duller than the second breakpoint's wave.

Within the sustain phase I then created a loop. During the sustain phase, the note changes its tone due to the differing harmonic compositions at each breakpoint. The loop takes a short phase of that constantly shifting tone and repeats it to mimic something of the character of an organ.

To then give more of an organ like quality to this sound I have also added a low frequency oscillator to give some vibrato.

As with the previous patch, I quite like the attack phase of the notes, but I am less sure about the sustain phase—I feel it could get quite wearing if you used it for a sustained pad. However, you should remember that these patches are intentionally made quickly—the main reasons for this was to show that additive can be a fast way of working and to demonstrate the basic principles. To really gain some benefit of additive synthesis you need to invest far more time than these patches would imply.

FM or Additive?

One question you may have is when do you use FM and when do you use additive? To my mind it is really a matter of what you feel easier with and how much control you want to have.

If you want to create a tone and to control how that tone changes over time and to be quite precise in the changes you are going to make, then an additive wave in Cameleon 5000 is likely to be the right answer.

On the other hand, if you want a tone that shifts over time, but you don't have the patience to adjust the wave at each breakpoint, then perhaps FM is the ideal starting point. However, if you are looking for an unusual waveform which may be layered and filtered, then perhaps an additive wave or two created in Rhino may be your ideal starting point.

As with all of these matters, the final decision is a matter of your taste and finding the tools that you feel comfortable using.

Chapter 10
FX

There are many aspects to consider when looking at FX in the context of sound design. There are several good reasons for using on-board FX and there are equally some good reasons not to use these units.

The FX units in most synthesizers are not of the same quality as commercially available plug-ins (although the quality is improving and this is not a universal damnation of all built-in FX units). By comparison, the built-in FX units are usually designed for a specific purpose and therefore don't have unnecessary features. Most built-in FX units usually work well with the synthesizer to which they are attached.

It is much easier to select an internal FX unit than to insert and route an external unit. Added to this, most built-in FX are coded with a lower hit on the CPU which is partly a reflection of their reduced feature sets. Then again, if you want CPU efficiency, it may be preferable to have ten synthesizer all feeding into one high quality/high CPU using reverb unit rather than have 10 synthesizers all running their own lower quality reverb unit.

Built-in FX units may not be as flexible, nor as controllable as some external plug-ins. However, built-in FX units cannot be configured in the same way as external units. As an example, with most synthesizers it is difficult to insert an EQ unit before a reverb, so that the EQ only affects the signal sent to the reverb unit and the dry signal is not affected by the EQ.

There are some FX units that are integral to a patch's design—for instance, distortion. Integral FX units allow a sound and the effect to be stored as one patch, making the sound more transportable. If external FX are used then each element would have to be separately stored if the sound is to be reproduced on a different system.

Lastly, you can do some really slick things with built in FX units that you cannot achieve with external units. For instance, you can:

■ control an FX send with velocity

- vary the depth on an effect with aftertouch

- "duck" an effect so that it is not heard while the played sound is heard.

One other personal thought—I've got to admit that, while I admire some people's ability to stick with an idea and explore every possibility with a set concept, I get pretty bored with all of the patches based on two slightly detuned sawtooth waves where the only tone shaping is achieved with the FX. I'm also of the opinion that a patch should be strong enough on its own without any FX—the FX should then be the added spice.

Why Use FX

FX units are generally used in two ways:

- as an integral part of the sound for instance, a distortion unit, or

- to enhance a sound, for instance, reverb to give a sound some space and depth.

The disadvantages of FX include:

- over use (often clichéd use)

- inappropriate use for a particular track

- excessive demands on the CPU, and

- muddying of the sound.

Deployment of FX

FX units are deployed in one of two ways: as insert FX units or as send FX units. Let me explain the practical differences.

Insert FX

With insert FX the whole of the audio signal passes through the FX unit (see Figure 10.1). So if you want to add a distortion effect, then you will want the whole signal to be distorted by the fuzz box—it would be rare (but not impossible) to mix a clean and distorted signal together when you have full control over the amount of the distortion. EQ and compression are other typical examples of insert FX.

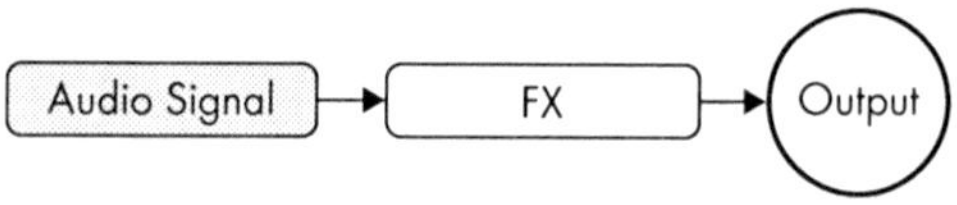

Figure 10.1 The signal flow through an insert FX unit.

Audio Signal → FX → Output

Send FX

With send FX, you send part of your signal to the FX unit and then add a purely effected signal back to the dry sound (see Figure 10.2). Examples of FX units that are grouped under the send FX heading are modulation effects (chorus etc), delays, and reverbs.

Figure 10.2 The signal flow through a send FX unit.

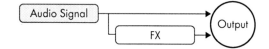

Using FX in practice

Most of the FX units in the six featured synthesizers work as insert FX. Generally they are inserted after the sound has been created (so after the oscillators, envelopes and filters) and before it reaches the output. For FX units that are commonly deployed as send FX (such as delays and reverbs) the FX units are still deployed as insert FX and the amount of the effect is controlled by a wet/dry mix control. Surge allows its FX units to work as send FX units to improve efficiency by sharing these units between two Scenes, however, these routing choices don't give you any additional sonic options.

FX units

All of the six synthesizers offer built in FX units—some have more, some less. Some have different flavors of similar units, others just give one option. Here is a brief summary of the generic units and an explanation of some of the uses you can put the FX units to.

Distortion

Most people are used to hearing distortion used in an extreme way. However, depending on the amount of control you are given over the distortion, it can be used in a much more subtle manner. Typically, you should be able to squeeze the following spectrum of tone color from most distortion units:

- At very low levels, you are unlikely to hear much effect, but the distortion can have the effect of warming up a sound and perhaps compressing it a bit.

- With a touch more effect, you will often find that a sound can feel more "vintage" as the distortion effect takes out some of the top end of the signal.

- As the effect increases, the signal will start to break up—this might be a useful setting for sound effects.

- The next step is to move into proper overdrive (start thinking of rock guitars and Deep Purple type organs and you will get the picture). This overdrive type effect can be quite controllable and quite musical.

- Finally you will get out-and-out distortion which may be hard to control or use with any subtlety.

All of the synthesizers featured in the book offer some sort of distortion (coupled with filtering):

- Vanguard takes the most straightforward approach, having a drive knob in the amplifier section.

- Rhino and Surge offer a single distortion unit. Rhino offers a fairly basic sort of distortion unit with the facility to balance the clean and distorted signals together. Surge also offers a single distortion unit, but this includes several tone shaping options.

- Cameleon 5000 offers slightly more options, including the choice between digital and analog-emulated distortion. One other useful feature within the Cameleon is a compressor (these are discussed in more detail below) and a control to thicken up the sound. Cameleon then offers the option to feed the distorted signal into a separate filter to smooth out the sound a bit more.

- Z3TA+ offers a choice of five distortion modes: soft drive, hard drive, valve amp, smart shaper and heavy metal. Each has a different character and is more suited for specific tasks. In addition to the different modes, there is also a sample rate reduction stage within the distortion unit. One other useful feature of the Z3TA+ distortion unit is the facility to route the distorted signal to a filter to further control the sound.

- Wusikstation offers eight different units, including multi-band distortion units which allow overdrive to be applied to different parts of the frequency spectrum.

When you use a distortion unit it is usually very easy to get a piercing sound which dominates the track, rather than a warm overdriven sound that may fit in context. If you are trying to recreate the sound of overdriven guitars, remember that all the rockers play their guitars through speakers—inefficient speakers that cut off most of the signal above 4kHz. To recreate this effect you may need to use a filter, some EQ or perhaps some form of speaker simulator. Z3TA+ offers a form of speaker simulator and Wusikstation has a Combo effect which emulates some of the properties of an overdriven amp and speaker combination.

Also, if you are looking for realism, then the distortion needs to be very controlled. Even heavily overdriven guitars can still sound clean (ish) if they are played very lightly. Ideally you should be able to control the amount of distortion by the velocity of the notes. To get more control and more "gearing" in the distortion (in other words, a much greater range of overload), modulate the distortion depth with velocity as well as controlling the volume of the signal with velocity.

One other thing you can do if you are trying to create a screaming lead guitar-type sound is add a high pitched sine wave to simulate feedback—give this sine wave a very slow attack (perhaps five seconds) or make it separately controllable (for instance, assign its level to the modulation wheel) and you will have another dimension of control.

Compression

Compressors were originally used to prevent a signal overloading an input, and indeed, compressors still are used for this purpose. In essence, a compressor is an automatic gain control: when a signal gets to a certain level (the threshold), the compressor restricts how much louder the signal can get (see Figure 10.3).

In practice, you're quite unlikely to want to use the compressor as a volume control and instead will use it as an effect where it can help to:

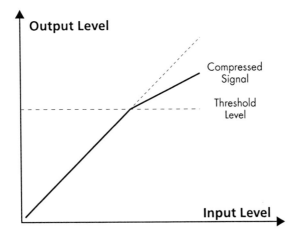

Figure 10.3 A compressor controlling an input signal.

- make a sound fatter and/or smoother

- enhance the perceived loudness of the sound, and

- to give the sound more punch.

The disadvantage of using a compressor is that the resulting sound can tend to dominate a mix. With more extreme compressor settings, the sound will usually have a more uniform level and will fill a broader frequency range. Both of these factors can make it harder to mix a highly compressed synthesizer sound. Another downside of compressor abuse is that the resulting sound tends to be less interesting—the delicate harmonic shifts of a patch tend to get covered up as all of the harmonics take a similar level.

Cameleon 5000, Wusikstation, and Z3TA+ all offer various forms of compressor. Surge offers a limiter (as part of its Conditioner unit).

EQ

The purpose of equalization is to make certain elements of the sound spectrum louder and other elements quieter. You can use EQ creatively or surgically. However, remember that you still need to get your sound to sit in a mix, so go carefully with the EQ. Make sure you are not using the EQ to try to cover the faults of poor programming.

The most flexible form of EQ is the parametric EQ (see Figure 10.4). With parametric EQ, you can set:

- the frequency to be boosted or cut,

- the amount of boost or cut, and

- the bandwidth of the boost or cut. The bandwidth determines how wide the boost or cut is, much like a constantly variable slope control for a filter.

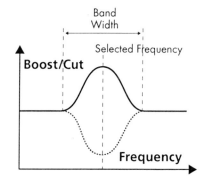

Figure 10.4 A parametric EQ unit.

A slightly more specialized form of EQ is the shelf EQ. A shelf EQ will cut or boost:

■   the whole signal above a specified frequency in the case of a high shelf (see Figure 10.5), or

■   the whole signal below a specified frequency in the case of a low shelf (see Figure 10.6).

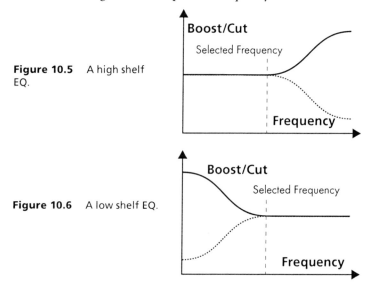

Figure 10.5 A high shelf EQ.

Figure 10.6 A low shelf EQ.

With shelf EQ you usually get control over the frequency (above or below which the signal is cut or boosted), and the amount of cut or boost. Sometimes you also get control over the slope in the same way that a parametric EQ give you control over the bandwidth.

A less flexible, but much quicker form of EQ is the graphic EQ. With these, the number of bands, the frequency of the bands, and the bandwidth is predetermined: you get control over the amount of cut or boost.

EQ sections are available in Rhino, Surge, Wusikstation and Z3TA+.

EQ in Rhino

Rhino has a three band EQ section, or rather three resonant filters which can cut high, mid, or low frequencies. The base range of the EQ is controlled by a cut-off slider which controls all three bands.

EQ in Surge

Surge is straightforward and offers three parametric EQs. However, since there are eight possible FX slots, the number of EQ bands is considerably more than three.

EQ in Wusikstation

Wusikstation provides seven four-band equalization sections—one for each of the six layers and a further master EQ section. Within each four band section:

- one band is a dedicated low shelf EQ which cuts or boosts all frequencies under the a specified frequency

- two bands are semi-parametric bell curve EQs which can boost or cut frequencies in a selected range, and

- the final EQ is a high shelf EQ which boosts or cuts the sound spectrum above a specified frequency.

For each of the three types of EQ, the bandwidth (or in the case of the shelf EQ, the slope) is preset (so only the frequency and the amount of cut or boost can be controlled).

EQ and Simul-Mode in Z3TA+

Z3TA+ includes a seven band graphic equalizer. The bandwidths are preset and there are 13 different options for the center frequencies that are controlled by the EQ section—this allows the equalizer to be more focused on different areas of the sound spectrum.

In addition, other FX units within Z3TA+ (such as the modulation effects and the delay) also have EQ included.

Within the Z3TA+ EQ section there is a simulator with 30 different amp/cabinet responses to provide equalization control that cannot be achieved with conventional EQ units.

Filter

As mentioned above, Cameleon 5000 also has a separate filter in the FX section. This is particularly useful when coupled with the distortion unit. None of the other synthesizers offers a regular filter in their FX section, however, Z3TA+ has the option to route the distortion signal back to the filter and Wusikstation has a simple filter stage in its master section as well as filter controls in some of its distortion units.

However, Rhino does offer the Crazy Comb effect which is based on two comb filters. This tends to give more of a chorus type effect rather than act as a filter.

Modulation Effects

These units (chorus/flanger/phaser/ensemble etc) all do a similar thing but the sound they produce can be markedly different. In essence they mix a delayed, modulated signal with the original signal to give a fuller/warmer/richer/whooshier (insert whatever description you want) sound.

Used sparingly, these effects can give an element of spark or richness to a sound. Used to excess and you will create a sound that quickly becomes passé.

Phaser/Flanger

Words cannot readily describe the difference between the sound produced by a phaser and the sound produced by a flanger, however, to my mind a phaser gives a more nasal sound where a flanger gives a more whooshy type sound.

Flangers and phasers are often used on percussion sounds and with plucked-type sounds (such as guitars and electric pianos). The sounds is very distinctive and is hard to use in a subtle way, so you are likely not to want to use the effect too frequently.

Chorus/Ensemble

Chorus and ensemble units also mix a delayed signal with the original signal, however, they use a longer delay than would be typical with a phaser or flanger.

The result can be a much more lush/smooth sound—at extremes however, it can sound like vibrato played too fast or just bad tuning.

Used in moderation, the effect can either add sparkle or thicken up a sound without noticeably changing the sound's characteristics. As an example, it is quite common to use a chorus with a bass guitar to provide a more rounded and brighter sound.

When used to the extreme you can get some serious thickening—you can also get a seriously mushy sound, so be careful if this is not your intention.

Rotary Speaker

A rotary speaker effect is intended to replicate the behavior of a rotary speaker. Only three of the synths—Rhino, Surge, and Wusikstation—offer this effect.

A General Word of Caution about all Modulation Effects

All modulation effects tend to have both a brightening and a thickening effect on a sound. This means that the effected sound can dominate the sound spectrum to a greater extent than the dry sound. Be judicious in your use of these effects if you are applying them to a sound that is intended to be used in the background—it may become more of a feature than you intended. If you are applying modulation effects to a background sound, then look at your arrangement and consider thinning out the arrangement to ensure the modulation effect does not become dominant.

Delay

All of the units offer some kind of delay. In each case, the delay time can be linked to the tempo of the track, being defined as a fraction of the beat (for instance, with delay of 1/4, the signal is delayed by a quarter note).

As you would expect, the featured synthesizers offer several types of delay.

Mono Delay

A mono delay adds a simple delay to the input signal. The only synth to offer a mono delay is Vanguard. However, a mono delay can be achieved in the other units if the delay time is equal for both channels in a stereo delay.

Stereo Delay

With a stereo delay, there are two separate delay lines which may (or may not) have different delay times. Conventionally the delay lines are hardwired left and right. All of the featured synths offer some form of stereo delay.

Cross Delay

Like a stereo delay, a cross delay has two channels. However, the output from the first delay is fed into the second and the output from the second is fed into the first. This can give the effect of spreading the delays across the stereo spectrum or ping-ponging the delays between the two channels.

Rhino, Surge, Vanguard, and Z3TA+ all offer some form of cross delay.

Other Common Variations

You may come across other variations of delay. These don't do much that is different from the three types of delay described above, instead they mix up the various elements, perhaps giving more delay lines or bouncing the output around the stereo spectrum in different ways. Z3TA+ offers a Ping Delay and an LRC Delay where Vanguard offers a Widen delay which slightly delays the right channel given a broader stereo spectrum, and Wusikstation offers a 3 Tap Stereo delay and a 3 Tap Stereo Parallel delay.

Common Delay Controls

The most common controls that all of the delay units have are:

- **Delay time.** As noted earlier, delay time is often (but not always) synchronized to the host's tempo and expressed in terms of beats.

- **Feedback.** Feedback controls how many times each delay repeats (quite literally, it controls how much of the delayed signal is sent back to the input).

- **EQ, damping or a filter.** A combination of EQ, damping, and a filter will control the sound of the echo. Typically these controls will be used to cut the high frequency elements of the delayed signal to make the sound duller, resulting in a more natural sounding echo.

Using Delay Lines

When adding delay, you need to consider what you are trying to achieve. If you are adding the slightest touch of slap-back just to give a bit of spark and a bit of space, then make sure you have done that. If you want to add a bit of echo, make sure that the delays are duller than the original source. If you are after an effect, then, as with the modulation effects, moderation is key.

Like modulation effects (and reverb), you also need to be cautious about how much space in the sound spectrum an echo will fill. There are several ways to tame the echo so that the sound doesn't dominate:

- Most obviously, by reducing the feedback and cutting the volume of the echo.

- Filtering (both high-pass and low-pass) the delayed signal so that the echo fills a narrower portion of the sound spectrum.

- "Ducking" so that the echo does not sound while the main sound source (that is being echoed) sounds—this effect was used conjunction with reverb in Bass Modulation (see Chapter 6: Modulation in Practice).

While an echo may dominate the sound spectrum, I often find that it dominates less than reverb while giving more character to the sound than reverb. I also find echo brighter and less mushy. For this reason I will often use echo rather than reverb when building patches. However, this is very much just a preference of mine. Ironically, I often then like to add some reverb to the delayed signal alone (so no reverb is applied to the dry signal) to smooth out the echoes and to move the echoes further backwards in the sound spectrum. Again, this is just a preference of mine—you may not like this technique.

Reverb

Reverb can give a sound some spatial context.

All of the featured synthesizers offer a reverb unit, but the quality varies from average to quite good. However, none of the reverb units offers as much control as a good external reverb (whether hardware or software) not least since you cannot put EQ before a reverb.

From a practical perspective, most reverb units use considerable system resources. If you are using (for example) six synths each with some reverb, it would be far more efficient to send all of the synths to a single external reverb unit. This may have the added bonus of giving a higher quality reverb.

All of the synthesizers featured in this book offer reverb—several offer different types and different algorithms.

With reverb units, there is significant interaction between the controls, so even more than normal it is necessary to balance several parameters when shaping the sound. The most common controls offered by the built in reverb units are:

- **Pre Delay.** This controls the time between the original signal sounding and the reverb taking effect. Smaller values give the perception of a smaller room while longer delays give the perception of larger rooms.

- **Size (or Room Size).** This controls the size of the room being simulated by the reverb unit. Usually this control has the same effect as a Decay control and affects the length of the reverberation (however, Cameleon 5000 offers separate Size and Decay controls).

- **Damp.** Damping works by cutting the higher frequencies in the reverb signal. One of the reasons why electronic reverb sounds artificial (that is clangy and metallic) is because it has too much high frequency information. The damping control can help to create a more realistic simulation.

- **Width.** Often controls the stereo width of the reverb.

Reverb in Rhino

Rhino gives three reverb programs—each has different controls and behaves in a slightly different manner.

- **Stereo Reverb.** Stereo Reverb is a quite conventional unit with fairly simple controls.

- **8-Tap Reverb.** 8-Tap Reverb bases its sound on two eight-tap delay lines. It gives a wider range of parameters (including pre-delay and diffusion) and gives a more natural reverb sound (to my mind).

- **Stereo Oktaverb.** Stereo Oktaverb is more of an effect rather than a reverb unit—it provides a limited reverb sound on the input signal raised an octave.

Reverb in Surge

Surge offers a single reverb unit, but it comes with a good range of controls, including some flexible EQ controls to help shape the tone of the reverb.

Reverb in Wusikstation

Wusikstation offers four different reverb units which offer various tone-shaping, and CPU consuming options.

Reverb in Z3TA+

Z3TA+ provides four different reverb algorithms, each of which combines a different grouping of parameters (such as pre-delay and reverb density) behind the scenes, but still allows the reverb to be edited by comparatively simple parameters. Each of the four algorithms has its own character suggested by the algorithm's name:

- Small Room

- Medium Hall

- Big Hall, and

- Plate—a simulation of plate reverb.

Pan

Only one of the synths includes a pan effect—Rhino with Autopan. This effect does what you would expect it to—it pans the signal between the left and right channels and gives you control over the panning.

A similar effect can be obtained with most of the other synths by using an LFO to modulate a signal's position in the stereo spectrum. The option in Rhino makes this effect easier to achieve (and is perhaps more controllable in certain circumstances).

Trancegate

This is one effect that has led to a series of imitators (or perhaps I'm too cynical and other developers had already had this idea, but didn't get to the market before reFX—I hasten to add that none of the imitators is featured in this book).

In concept the Trancegate is quite simple, see Figure 10.7. When the effect is switched on, a 16 step gate is engaged. The gate is directly linked to the tempo of the track with each gate step being a division of the beat, so for instance, if the speed is set to 1/16, each step will last for 1/16th of a beat and the trance gate cycle will last one bar.

Figure 10.7 Vanguard's Trancegate.

The audio signal passes through the 16 steps—at each step, the gate is either open or closed. This creates rhythmic patterns to the sound which still continues to progress through its cycle (as determined by the envelopes and other modulators). Vanguard offers separate gates for the left and right channels which further increase the options for this tool.

Z3TA+ takes a different approach to gating—when the note C0 is struck, the sound is gated. This means that the gate pattern can be constantly shifted or controlled in real time.

Chapter 11

Putting it All Together

In the next chapter I will look at some practical examples of sound design, but before we get there I want give something of an overview of the whole process.

Design Philosophy

The comments below are my personal opinions and should not be taken as facts. If you disagree with me, then your opinion is probably correct—I cannot know the context in which you will be using your sounds and so it would be presumptive of me to suggest that I could ever know better. So if you disagree, that is fine with me, but please don't write and tell me.

Programming with a Purpose

There may be many reasons to design a sound. For me the most pressing reason for building a patch is the need to find a specific sound that works for a specific track (or a specific sound that is *needed* in some other specific context). If you stumble across a sound that you think is interesting, it doesn't really mean much unless you can then use that sound appropriately. While you may program "just for fun", the remainder of the book is predicated on the assumption that you will have a specific goal when creating any sound.

Only you know how and where a sound is going to be used. And while it is something of a chicken and egg conundrum, I always find that it is more difficult to program the nuances of a sound without having a rough MIDI part playing. In practice, if you separate the arrangement of the part from the programming of the sound, you will always be at a disadvantage.

By equal measure, I would encourage you to try sounds in a different context from the one for which they were originally intended. For instance, many bass patches make great stab sounds. Don't do this as a last ditch measure when nothing else works, but rather as an exercise to find how different sounds can fit in a different context.

Many people will also encourage you to "just experiment" with programming. Personally, I dislike the notion behind this comment. I'm not saying that experimentation is bad or that you shouldn't try things. My only argument here is that the "just experiment" philosophy is used by those who can't be bothered to learn how to use a tool properly. Equally this argument is used by developers who can't be bothered to explain to people how to use a tool properly, or who have developed a tool which may be interesting but is ultimately without purpose.

Arrangement of the Track

There are many ways to get to a sound—use a preset, use a commercially available sound bank, or even hit the random button until you find something interesting. All of these are valid ways of getting a sound, so why would you bother programming your own sound?

As I have already mentioned, there is one simple reason why you would program your own sounds—and within the scope of programming, I would include tweaking a sound from another source—to get the *right* sound for your production.

When you are mixing a track, each instrument needs its own space or else the mix may get be cluttered, indistinct, and generally bad. You may have the best bass, pad, stab, and lead sound available in your synth, but do these elements all fit together? Does your bass muffle the kick drum? Do the bass and the stab occupy different sonic ranges? Does the stab stand out over the pad and does it then clash with the vocal?

These are all elementary problems associated with the mix which are often addressed with equalization.

Even if the sonic ranges occupied by your synthesizer parts aren't the problem, have you found the right sound for the track? Does the bass fit that pumping rhythm you've worked so hard at? Does that lead cut the top of your head off, or is it just a flaccid preset that seemed alright when you sequenced the riff? Does that stab really lock in with the bass to give a perfect performance or is it an unequal match for that bass? And what about that pad? Does it sound like something from a 1970s string machine? Or worse still, does it sound like a cheap VSTi when you want it to sound like a 1970s string machine?

Automation is an important part of the mixing process, so why not also consider automation in the production of your sound? Don't just automate a patch's level, but think how the timbre can change over time—control this with player controls and with track automation (for instance, open and close the filter slightly as the track plays).

Can't I Just Go and Buy Something?

A simple question that is often asked is "what is wrong with commercial banks and the banks that come with my synthesizer?"

As simple question deserves a simple answer: there is nothing wrong with presets. The only downside to presets is that they're just not designed to fit your specific track and you may not know how to control them: for instance you may sweep the filter when it may be more appropriate for the sound you are after to sweep the depth of the envelope modulation of the filter.

People (including me) use presets for many reasons—some are excellent, they save programming time (so you can make music), they sound good, the results are better than you can program yourself, the style is outside of your usual programming style, or the presets give you inspiration. As I said, I use presets and will continue to use presets in addition to programming my own sounds.

However, it seems that you're being very hopeful that you could take a sound that has been programmed in a vacuum (in other words, the sound was programmed by someone who knows nothing about your track), *and* that the sound will then be used in a context of a mix, *and* will work perfectly. I can believe that a patch programmed out of context might be right for your track—I find it hard to conceive that it would fit the mix perfectly *and* would fit the playing style of the MIDI track where it will be responding to velocity and other playing techniques.

I think you also need to consider the purpose of factory presets and what sound designers have to do to persuade people to purchase a bank of patches. Presets that come with a synth have one purpose—to display the capabilities of that synth. This is a sensible thing to do and is exactly what I would do if I were in the business of making synths.

However good the presets are, however well they demonstrate the synth in question, they are not necessarily right for your track. So what do you do? Most people consider buying some presets from somewhere else. Which presets do they choose? Usually the one that sounds most flashy—not necessarily the one that fits with the track they are trying to construct.

Some (but not all) commercial banks are not very playable (for instance they don't react to velocity or other playing nuances and they don't have controllers set up). They are also often drenched with effects (where this may not be wise/appropriate)—for instance you will often get a bass swimming in reverb. This may sound good when one or two notes are tried, however, in the context of a track it often just sounds muddy.

Don't get me wrong. Some commercially available presets are genuinely inspiring—especially some of the more modern dance focused banks made by guys who are actually making modern dance music. These are patches that have been designed for a purpose by people who actually understand how to deploy the patches—that is why their demos sound so good.

If you have a choice between bad programming done by yourself and a good preset, then the good preset should win every time. Sound programming should *never* detract from the music and the music making process. You should also remember that there is no rule that the sound design has to be undertaken by the creator/producer of a composition: it is a perfectly valid decision to use external sound design, provided the result works for the track.

Finally, please don't form the view that I think you have to program your own sounds to create a musically valid piece. I'm just as critical of the use of presets you have created yourself if they are inappropriate for a track.

Knowing the Limitations

I have made many references to real instrument sounds. This has been done because most people have an idea of a "piano sound" or a "brass sound". However, I would not recommend using any of the techniques in this book if your goal is to capture the nuances of a real instrument.

If you are after authenticity, then I recommend you hire a studio (with an good room and an engineer), hire the instrument you want to record, and find the best session musician you can get. This is most likely to give you satisfactory results. As a second option, find a quality sample library and use samples of the real instrument—again, find a session musician as they are likely to give you the best performance.

That being said, don't be shy of using synthesizers—just remember, they are not the real thing (unless your goal is synthetic sounds).

Polyphony Limit and Note Priority

Any machine you use to create music is going to have limitations. This is true whether you are using hardware or software. There will always be a finite number of notes you can play. With hardware, this limit is usually a known number (often hardware will be described as being, say, 64 note polyphonic). With software, the system limitations are usually affected by a range of factors, such as how much system resource is used by other functions (such as the host, the effects in the synthesizer, the number of filters you are using etc).

With hardware synthesizers the limitations on voices may lead to voice stealing—when you reach the maximum polyphony, the synthesizer will cut off a previous note as a new note is triggered. With software, when the system resources are reached then your system is likely to drop the audio, freeze, or fall over.

However, there are many things you can do with software to reduce the load of your CPU.

Freeze Tracks

Most hosts will allow you to freeze tracks and free up resources. This may be the easiest way to prevent system overload.

Limit Polyphony

With software you have the option to limit polyphony. By limiting polyphony you will have to address voice stealing rather than look at complete system failure (which is probably the less preferable option). As the voice stealing can then be performed on a per patch basis, you have the option to control the voice allocation so that key parts are not robbed.

All of the featured synthesizers allow you to set the polyphony limits on a per patch basis. With Wusikstation you can even limit voices on a per layer basis and so ensure that the more important elements of your sound are not compromised. Cameleon 5000 also allows you to limit the number of partials used, however, this may have an effect on the tone.

Simplify the Patch

With software there are many things you can do to simplify a patch and reduce its CPU load, for instance:

- **Reduce the number of oscillators.** Is that sixth oscillator really adding to the sound or is it just there for good luck? Equally, as we mentioned in Chapter 4: Sound Sources, it is possible to use one wave where it creates the sound of two waves in combination.

- **Reduce the filters.** Either reduce the number of filters used and/or reduce the slope of the filters (for instance in Z3TA+ the 36 dB filters use more system resource than the 12 dB filters).

- **Cut the release time of your envelopes.** A shorter release time will mean that the total number of notes still sounding at any one time may be reduced. This can be a very effective way to reduce total system load without losing polyphony especially with very busy playing. However, it may make your patch sound awful.

- **Turn off the FX units.** As noted in Chapter 10: FX it can be a more efficient use of system resources to use external FX units.

Principles of Sound Design

Unfortunately there is no magic formula to creating a patch. However, I would suggest a few basic steps before you jump in:

- Understand what each knob does.

- Understand the signal flow in the synthesizer and how each element interacts.

- Have a definite aim in what you're trying to program—don't just fiddle with the knobs and hope something useful comes along.

Once you have these basics sorted, I suggest a two stage approach:

- **Get something that is "alright".** In other words, a sound that is functional and is reasonably close to where you want to end up—you should be able to get to this point quickly. Then

- **Undertake the detailed tweaking.** This is the time to make sure the patch is exactly right for your needs—this will be the time consuming part of the programming.

When it comes to getting something that is "nearly" right, I see no reason not to use a preset, although the subsequent editing may take longer as you will have to familiarize yourself with the workings of the patch. But then again if you're using one of the synthesizers with a more straightforward architecture or you know what you're doing, you may be happier starting from a clean sheet.

So you know how your synthesizer works, you have a grasp of what all the knobs do, how do you get the basic sound together. For me, I then tend to think in terms of the main food groups.

Main Food Groups

Most sounds can be characterized by two key elements—their brightness and the envelope (primarily the attack of the envelope, but often you need the decay and sustain level to be right for the contrast of the attack to be clear). Once these two elements have been addressed, any tonal nuances after that may be affected by a range of factors.

Brightness

If you are trying to categorise the brightness of a sound, it will generally fall within one of four descriptions:

- very bright
- bright
- dull, or
- very dull.

You can try to subdivide further, but realistically, it is already quite difficult to try to find the line between these four categories.

Attack

If you try hard, you can find four categories of attack time:

- fast—percussive (for instance a xylophone hit or a guitar string being plucked)
- fast—but not percussive, for instance an organ
- medium, and
- slow

Again, you can try to subdivide the categories further, but it is a fairly pointless exercise.

Getting the Combination Right

Once you understand what you want in the way of brightness and envelope, getting the right sound is dependent on the right elements being drawn together. For instance, here are some combinations may want to look at.

To Get Brightness

Consider the oscillator (or oscillators) and filter combination. If they don't work together, you will either get a shrill sound, or a thin sound, or a dull sound, but never a bright sound. For instance, if you are creating a sound based on a square wave, you may find it preferable to choose a 12 dB filter. A 24 dB or 36 dB filter may work, but to my ear they ultimately rob the square wave of much of its character, which you may want to keep if you are trying to create a bright patch.

To Get Richness

One of the key elements to a rich sound is its movement. You can achieve this with careful selection of your sound source (which will usually involve several slightly detuned oscillators and perhaps some delicate use of a sub-oscillator). Alternatively, you can just slap on loads of chorus. My preference is for the former option.

To Get Attack

Most obviously, it is important that the envelope opens quickly to get some attack in a note. Depending on the tonal quality you are after, it is often important that the note then decays quickly to the sustain level—this will give you a percussive envelope reminiscent of a plucked guitar string or similar.

However, to ensure that the attack is emphasized, another significant factor is the waveform. Some waves can sound slower than others. For instance, all other factors being equal, a square wave can sound like it has a faster attack than a sawtooth wave. In general bright sounds (especially distorted bright sounds) are perceived as having more attack.

To Get Warmth

A warm sound is subtly different from a rich sound, often it is much less bright. Warmth is also a comparative term—if a colder sound starts a patch and is then washed over by a warm sound, the contrast can give the listener the illusion of warmth.

Any warm sound will probably have a slow attack and a slow release. There must be thickness to the sound which will often come from detuned oscillators. As noted above, one of the key elements of a warm sound is the changing tone which can be achieved with another oscillator with an even slower envelope.

To Get Punch

It is hard to get some punch into a sound without adding too much brightness. First you need a fat sound source, then you need a fast attack and finally you need a hefty filter. The filter can be modulated with the envelope so that the attack is emphasized but the brightness is controlled.

Choosing the Right Tools

When you are programming take some time to make sure you choose the right tool. It is easy to make a few glib comments about which is the right tool for any job. For instance if you want a fat "analog" sounding pad, then perhaps your first choice should be Vanguard—that will likely give you the fastest and the best results.

However, Vanguard is not the only tool you can choose—Rhino makes excellent FM sounds, but in the hands of a skilled sound designer it is equally capable of making great analog sounding patches. Which is right for you is a matter of your taste, your skill, and what is right for the particular track you are working on. However, don't let your (or my) preconceptions about any machine's strength let you rule out a synthesizer from any task.

More importantly, choosing the right tool is more a function of having the knowledge to control that instrument. The only way to learn how to use an instrument is to keep using it over a long time—don't keep buying new instruments, pick one and learn every aspect of its character.

Playability

Finally, any patch has to be highly playable. The playability of the patch is one aspect that separates the average sound from the professional. Use your knowledge of how the sound will be deployed to set the modulation sources.

For instance, if the sound is going to be played by a live musician, then you might ensure that velocity has an effect and that the modulation wheel and expression pedal both offer a wide range of control. However, if the MIDI track controlling the sound is going to be programmed, then you might want to make sure that external controllers, such as track envelopes may have a significant effect, so for instance, you may add an additional filter which has little effect in the normal course but can be readily controlled by a track envelope.

Chapter 12
Building Patches

This chapter is broken into five pieces: this first part and the four subsequent refills. This original chapter was included when the book was first published. The refills were added when the book was updated and re-published in hard copy form in 2007.

So far we have designed some sounds, but many of these patches have been used to illustrate specific points. This chapter focuses on creating sounds that are usable in a musical context and which give the musician a lot of control over the sound.

One of the best ways to learn about sound design is to look at existing patches and take them apart piece by piece. However, this may not always be the most straightforward way to learn. To draw a parallel, you don't learn to do jigsaw puzzles by taking completed jigsaw puzzles apart. For the same reason, reverse engineering patches may not always be the easiest way to learn how to program. In my view, you learn how to program by programming, and with experience you improve your skills.

That being said, once you have worked your way through this chapter I would encourage you to dissect other patches to understand how other sound designers create their sounds. Then try to recreate those sounds and learn from your mistakes.

So far in this book many of the example patches have been fairly dry, lacking effects. For this chapter I will be adding effects. However, my taste is (generally) to use less rather than more and as I have mentioned there are many reasons why you would want to rely less on built in effects and instead use master effects in your host (for instance to save CPU usage and to use higher quality effects).

Accordingly, in this chapter you may feel that I have either used too much or to little in the way of effects. As always, there is the question of taste—these patches have been designed to my taste and for the uses I want to put the patches to: your tastes will be different from mine and so you may find some of these patches more, or less, useful.

As I mentioned, the patches in this chapter are all intended to be used in musical situations. However, if you have the banks you will see that I have added a few patches to help illustrate how I started to build particular sounds. These patches will be mentioned, but will not be reconstructed. Lastly, there are also one or two patches mentioned (but again not reconstructed) where the result was interesting, but was not the sound I intended. These are included so that you can learn by my mistakes.

One other point—just because I have demonstrated a patch with one synth, there is nothing to stop you apply the principles with another synth although you may have some difficulties translating some of the FM principles to all of the other synths. For some patches you will also have to make substitutions—for instance, only Z3TA+ has a 36 dB filter and a Six-Voice Chorus. These may make some patches harder, but not impossible, to replicate. Perhaps you could use different filters/FX etc and take the patches in a different direction.

If you've got this far and are still reading, I would highly recommend you get hold of the patches that accompany this book—it will make this chapter far easier for you.

Basses

The first group of patches I am going to look at are bass sounds.

Z3TA+ Bass Patches

Z3TA+ is a great synthesizer for making bass sounds.

1970s Bass

For this first Z3TA+ patch I wanted to build a simple fat synth bass sound which is reminiscent of the synth bass sounds that were popular in the 1970s. I chose to make this patch in Z3TA+ as it gave a thick and rich sound. I did try this patch in Vanguard (listen to Slider), but couldn't get the effect I was after (in terms of sound) and also Vanguard didn't give me the control over the filter envelope that I was looking for.

The heart of the sound is created by oscillator one and oscillator two (both key synchronized). Oscillator one uses Vintage Saw One and oscillator two uses Vintage Square One, but dropped an octave below oscillator one. The level of oscillator one was set at 75%—oscillator two was then increased until it had a thickening effect on oscillator one, but not so much that it overpowered the sound.

When the patch was finished, the final audio test I did to balance the Level for oscillator two was to ensure it was loud enough to add weight to the attack portion of the sound, but quiet enough that it had little effect during the sustain portion of the note. To my ears, setting the Level of oscillator two at 32% gave the right balance.

Once oscillator two had been sorted, I added oscillator three to double oscillator one. To thicken the sound, I slightly detuned oscillator three. The effect of these three oscillators together is very bright and very dominating. The sound is also a bit fizzy and lacking any control by the player.

The next step was to engage the filter. For this I called up the 24 dB low-pass filter. The filter block does two things in this patch:

■ first, it shapes the tone of the patch, and

■ second, it controls the level of the patch.

The tone is controlled by envelope one and the velocity, but the volume is controlled by the velocity alone. As I'm only using one filter, I set the oscillator busses to send 100% of their signal to filter one (bus one).

The interaction between the modulation matrix and the filter is (to my mind) slightly counter-intuitive in Z3TA+ unless you are using the curves. The filter cut-off slider determines the level *from* which the filter is cut-off is modulated. Whereas the filter level slider determines the level *to* which the filter level is modulated. From a practical perspective I can understand the different operations, however, when setting up the modulation matrix you need to be aware of that difference.

To set the filter for this patch, I put the Cut-Off slider to zero so that the cut-off frequency would be fully controlled through the modulation matrix. In the matrix, one row controls the filter:

■ the modulation Source is envelope one

■ the Control is Velocity (so the extent to which the filter cut-off is modulated by the envelope—between the minimum and the maximum set by the Range—is controlled by velocity), and

■ the Destination is filter one cut-off, as you would expect.

There is a balance to be struck when setting the Sustain Level and the top end of the range control. My approach is to set the Sustain Level to a place where it is too low and then to tweak the range control. This allows me to adjust the point to which the envelope's attack opens the filter to the maximum extent. In this case, I was listening for a sound that wasn't too bright.

Once the maximum extent of the range has been set I go back and adjust the Sustain Level of the envelope to get the brightness (or dullness) during the sustain portion of the envelope right. With this patch I disengaged the velocity control of the modulation matrix when setting the maximum extent of the range.

For this patch, setting the Maximum in the Range control to 80% sounded right to me. I then re-engaged the velocity control in the modulation matrix and set the Minimum in the Range. In this case 25% sounded right to me.

However, although the velocity gave me the control I wanted over the filter, there were still two problems:

■ first, the sound was still too thick, and

■ second, the filter envelope was wrong, which wasn't surprising as the only adjustment I had made to the envelope was to set the Sustain Level to be too low.

The thickness of the sound was rectified by pushing up the Resonance slider in the filter. At around 9.44 dB this gave the right sound and also emphasized the effect of the variable filter cut-off. In setting the filter envelope I wanted to do two things:

- first emphasize the attack of the note—this meant that the level of the envelope had to drop swiftly after the attack portion of the envelope, and

- then have the envelope continue to decay gently over time so the filter continues to close as the note is held.

To achieve this effect, I needed to use one of Z3TA+'s multi-stage envelopes. As the modulation matrix had been set up, all I needed to do was to set the envelope—this was done by ear: I listened until is sounded right.

In this case, the Attack was set to its fastest (zero). To give the bite to the attack, I set the Slope Level to 75% and the Slope Time (that is the time it takes the envelope to go from the maximum at the end of the attack phase to the Slope Level) to 0.05 seconds. This gave some bite to the sound.

So that the sound would progressively continue to lose its brightness I set the Sustain Level below the Slope Level. In this case I set it to 62.5% and the Decay Time to 5.2 seconds. This means that the note will not continue to lose its brightness after is reaches the Sustain Level. However, in practice, it is unlikely to continue to be held for much after this point so I am not too concerned about this.

With the filter we have done two things here. Firstly, we have set the tone of the patch. However, there is a second effect—the sound of the patch now fills up far less of the sound spectrum: it is still a big bass sound, however, now the patch clearly works as a bass sound and does not dominate areas outside of the bass range. That being said, you will still need to ensure that it works in the low end of the spectrum with the kick drum and any other sound elements in that region.

Linking the filter cut-off to velocity does give velocity some control over the volume. However, I wanted more and so I set up another line in the modulation matrix. This time the source was On, the Control was velocity and the Destination filter one level. I didn't want the volume to go from zero to maximum, instead I was looking for a slight reduction at lower velocities. To achieve this I set the Range Minimum and Maximum in the modulation matrix to 70% and 100%.

With the level of filter one set to 75% the patch has a tendency to overload. Ideally the level of the filter (or the oscillators) would be turned down until the overloading is removed. However, I'm lazy so I engaged the master Limiter. You should be aware that this uses additional CPU resources and so may not be the most efficient programming technique.

If this patch had been programmed in a synthesizer from the 1970s, the synth would almost certainly have been monophonic. Accordingly for this patch I set the polyphony to one and added some portamento. The portamento works so that when one note is held the next note slides to its pitch: there is no retriggering. The effect is sufficiently subtle that is can be heard for small intervals—at larger intervals the effect is not perceived. Finally, staying with the 1970s theme, no FX are added to the sound.

Mwuah

This next bass patch is based on a Vintage Sawtooth One wave dropped an octave and then fed through a 36 dB low-pass filter with the Cut-Off set to 983 Hz and the Resonance set to 7.43 dB. This gives quite a dull sound—for this patch I want to create a "slow" bass sound: as you can hear, this oscillator/filter set up gives us a fast and uninteresting sound.

The first thing I wanted for this patch was for the pitch to rise slightly when the note is struck. To do this I used the Pitch Envelope and set the Start Level to -8.75% and the Attack Time to 1.79 seconds. Using the Pitch Envelope as the Source to modulate oscillator one and having a maximum Range selected, when the Pitch 1O curve is selected, the effect of the pitch rising when the note is first struck can be heard.

The second tweak was to open the filter slightly when the note is struck and then close it very slowly. To do this I called up envelope two to modulate the filter cut-off—in the modulation matrix, the maximum Range was set to 6.8%. There are two stages to the envelope—the Attack Time which is set to 0.75 seconds and the Slope Time which is set to 10 seconds. At the end of the Slope Time, the Slope Level is set to zero (so the envelope has no further effect on the filter).

So far, the sound is not particularly interesting and it is too fast for a slow bass sound.

The third change was to separate the filters. There is no separation slider—this change has to be effected through the modulation matrix. The Filter Separation Destination in the modulation matrix is controlled by envelope one acting as the Source. The Range was set to the full extent and the Curve selected was U-Lin- in other words, at zero modulation the filters are separated to the maximum extent, while at maximum modulation there is no separation (so the filter acts as a conventional 36 dB filter.

The patch illustrates filter separation in practice—or more accurately, it illustrates recombining separated filters. The 36 dB filter is actually three filters stacked together. Each filter has its own cut-off and resonant peaks: in normal mode these resonant peaks are stacked and so are quite prominent.

With separation, the cut-off frequencies of the filters are spaced apart so successive cut-off frequencies are progressively raised. As the filters are separated, there are still three filters each with their own resonant peaks, however the peaks are at different frequencies and so are less prominent.

The net effect of separating the filters is a change in tone without too much of a significant change in the brightness. This gives the sound more of a vocal quality. Try playing the patch in higher keyboard regions to hear this effect more clearly.

Envelope one, which controls the separation, takes advantage of Z3TA+'s multi-point envelopes. The Attack Time was set to 0.26 seconds and having reached its peak, the level then falls over a Slope Time of 0.34 seconds to a Slope Level of 80%. This portion of the envelope gives the patch its characteristic "m-wow" sound. After this the level of the envelope rapidly decays to a Sustain Level of zero over a Decay Time of 0.95 seconds. At this Sustain Level, in the higher ranges of the keyboard, the filter is only kept open by the effect of envelope two.

So there you have it, a slow bass which in the higher keyboard regions gives quite a vocal quality. I like the sound as it is and so I decided not to add any FX.

You will notice that there are no player controls here. This is quite a slow patch and there was only one thing I wanted to do, however, unfortunately Z3TA+ doesn't allow it. Ideally I would like to modulate the Attack Time in envelope one (with velocity) so that the time over which the filters are recombined could be controlled. As I said, you can't do this in Z3TA+ and so there are no player controls here.

Bells and Chimes

I now want to look at how to make bell and chime sounds.

Z3TA+ Bells

With its FM sounds, Z3TA+ can make some great bell sounds. However, the synthesizer is much more flexible than that, as we shall see.

Bells

It is quite possible to create bell-like tones without FM. I quite like the non-FM way of creating bell sounds because the timbre is less sharp than with FM and there is almost a wooden tone to the sound in certain key ranges. As this patch is created with only two of Z3TA+'s six oscillators, it would be possible to layer some FM bell sounds to give a hybrid sound. Personally, I think this sound is good on its own and so I haven't followed that option.

This bell sound is based on two multi triangle waves—one is clean and the other is put through a filter with lots of resonance. In the patch oscillator two is fed to bus two (passing to filter two)—as no filter is selected, the signal passes through without being affected except where the level of oscillator two (which was set to 100%) is modulated by envelope two.

Oscillator one is fed into bus one and passes through the 24 dB filter. The cut-off frequency was set to 4,800 Hz and the Resonance set to 26.88 dB (the maximum). The Level of this oscillator was set to 85% and this level is modulated by envelope one.

The filtered wave has a bright metallic tone with an almost breathy quality. The unfiltered wave is less bright, but bringing it together with the filtered wave gives a warmer, ringing tone. However, there is no volume control and the percussive attack which is so characteristic of a bell is missing.

Both envelopes have a fast decay although envelope two has faster decay. Both envelopes also use Z3TA+'s multi-point envelope. Envelope one falls to the Slope Level (of 67%) over 0.13 seconds and then decays to the Sustain Level (of 17%) over 0.58 seconds. Envelope two falls to the Slope Level (of 30%) over 0.45 seconds and then decays to the Sustain Level (of 26%) over 0.1 seconds. The combined effect of the envelopes takes a metallic sound and gives it a bell-like quality.

To achieve the final sound, three effects units were added:

- the Six-Voice Chorus which adds some "shimmer" to the sound

- one Ping Delay which gives the initial attack of the bell more emphasis

- the Large Hall reverb which, in addition to adding reverb, also smoothes the sound giving a more coherent feel.

Keys

I now want to look at keyboard-type sounds.

Wusikstation Piano Patches

The first keyboard patches are going to be demonstrated with Wusikstation.

Piano

Piano is a comparatively simple piano patch. The basis of the patch is the Piano Grand waveform (from the Famous Keys Volume Four bank) loaded into layers one and three. These have then panned hard left and hard right.

The output volume of each layer has been linked to pitch by selecting Pitch as the modulation Source and the layer's Volume as the Destination. As the pitch increases, layer three's volume increases, where layer one's volume decreases. The effect of these volume changes is to spread the piano across the stereo spectrum so that lower notes are heard in the left channel and higher notes are heard in the right channel.

It would have been possible to use one layer with pitch as a source and pan as a destination. However, I preferred the sound created by having two layers.

For both waveforms, the layer envelope was set with a zero Attack and Decay time, the Sustain was set to its maximum and the Release time to around 60. The Velocity was set to 86 giving quite a responsive patch.

Each waveform passes through a 2-pole (12 dB) filter which is fully closed (with no resonance). The filters are both controlled by Modulation Envelope One. With the envelope, the Attack time was set to zero, the Decay time to its maximum, and the Sustain level to zero. The Vel knob was set to 76. The effect of the envelope is to introduce velocity control to the filter, and it also allows the brightness to fade as a note decays.

While interesting, to my mind, the samples need a bit of weight. The first bit of weight that I added was layer two with the Pure Sine wave selected. Unlike the other two waves, this decays over time—the layer envelope was set with an Attack time of zero, a Decay time of 94, and Sustain level of zero. The Velocity knob was set to 90.

The final bit of weight I added was in layer four where I called up the 202 Vel Bass 01 wave, again from the Famous Keys Volume Four bank. The layer envelope's Attack time was set to zero, the Decay time to 87, the Sustain level to zero and the Vel control to 127. The layer's Volume was then set to 61 and in the modulation matrix, Pitch was set as a modulator negatively controlling the volume (so the layer gets quieter at higher key ranges).

To complete the patch, layers one and three were then sent to the Lite Verb reverb unit.

Electric Piano

Much like Piano, Electric Piano is a very fast, effective piano patch in Wusikstation. In many ways this patch is probably simpler.

In layer one the Rhodes 73 wave is loaded and in layer two the Rhodes Vel wave is loaded. Both waves come from the Famous Keys Volume Four bank. The velocity range for layer one was set from 0 to 116 and for layer two from 117 to 127. Layer one runs through a 4-pole (24 dB) filter which is controlled by velocity.

Mod LFO One was set to 1/8 and introduces vibrato (when the modulation wheel is pushed up). In the modulation matrix, the Amount (of modulation) was set to 62: much more and the vibrato would sound unnatural.

Once set up, FX were added to the patch. In this case I added a Quad Chorus and a short echo set to 1/16 with a small amount of feedback and high damping, so the echo is acting more like a reverb unit.

The final tweak was to control how much signal was sent to the Quad Chorus unit—the idea was to have more chorusing at lower velocities but less at higher velocities so the aggressive tone could shine. This took two lines in the modulation matrix—one for each layer. The Source was Velocity and the Destination layer one or layer two FX1. The Minimum for each line was set to 113 and the Maximum to 31 with an amount setting of 127. To my ear, this gives the appropriate scaling.

Z3TA+ Piano Patches

One of the key themes I have emphasized is programming with a purpose. However, I stumbled on this patch when I was trying to do something totally different. From an interesting sound I then developed the patch into a highly usable Wurlitzer-type electric piano sound.

The basis of the sound is contained in Z3TA+ Piano Root. Those of you with the patches can look at this patch in more detail. In essence, the sound is created by oscillator one acting as modulator for oscillator two (the carrier). This simple FM stack creates a tone which can be changed in a very controllable manner (in this case by velocity controlling the output of oscillator one).

Oscillator two is then doubled with oscillator three (which is not frequency modulated) and the resulting sound is then fed through a low-pass filter. This architecture means that the tone is changed in two ways:

- the sound source changes (quite radically) by velocity controlling the amount by which oscillator one frequency modulates oscillator two, and

- the filter's cut-off frequency is also modulated.

Z3TA+ Piano takes this simple idea and builds a usable patch. We will use a similar technique with Male Voice Choir in the pads section, below.

Z3TA+ Piano

Z3TA+ Piano takes Z3TA+ Piano Root as its basis, makes a few tweaks and then builds. Again, the essence of the sound is oscillator one working as the frequency modulator of oscillator two which acts as the carrier. Oscillator three then doubles oscillator two (but is not frequency modulated so it adds much more to the weight of the tone than simply thickening it through doubling).

As with the earlier patch, oscillators two and three use the Vintage Saw One wave. However, in a change from the earlier patch, oscillator one uses Vintage Square One rather than the sine wave as the modulator. The change is subtle, but to my ears it gives the sound a bit more bite. In Chapter 7: Frequency Modulation Synthesis, I expressed the opinion that it is probably best to stick with sine waves for FM—I still hold with this view, however, in this patch there is extensive use of the filter and I feel that the change in modulator wave adds to the tone of the patch.

The two sound generating oscillators are then fed into a 36 dB low-pass filter (filter one) with the Cut-Off set at 540 Hz and the Resonance to 1.26 dB. This has the effect of robbing the sound of any real tone and so to rectify this and give a bit more shape to the sound, the filter is modulated by envelope one—this modulation is controlled by velocity and the Range is set to a Minimum of zero and a Maximum of 70%.

Envelope one, which controls the filter, has its Attack Time set to zero, the Slope Time to 0.11 seconds and the Slope Level to 43%. This gives a brightness to mirror the hammer strike of an electric piano—the envelope then decays over 4.5 seconds to a zero Sustain Level. Finally the Curve I selected was U-Lin+ as this gave the smoothest control over the widest usable range of tones.

The envelope is essentially a simple ADSR envelope—the second part of the envelope (the decay time and sustain level) allows the filter to be used largely as a volume control so that a note does not sustain indefinitely (however, this element of the envelope does change the tone too).

The reason for the simplicity in the filter is that much of the tone of this patch is created by the frequency modulation of oscillator two by oscillator one. The output level of oscillator one has a modulation source of envelope two. In the modulation matrix, this modulation source is controlled by velocity.

The output level of oscillator one is quite low at 25%, however, this is sufficient to give some real bite. As mentioned this output level is modulated by envelope two where the Attack Time is set to zero, the Slope Time to 0.66 seconds and the Slope Level to 75%. This gives an effect similar to a hold stage in an envelope and emphasizes the attack of the note—as the slope time in this envelope is longer than the slope time in envelope one which controls the filter, this gives a longer emphasis to the attack while the tone is controlled by the filter. Having reached the Slope Level, the envelope then falls to the Sustain Level 31% over 6.5 seconds.

So far we have created a serviceable electric piano type patch. However, I wanted to make a very playable patch, so this is what I did.

The first thing I wanted to do was to fatten up the patch. This was quite simple—I called up oscillators five and six, selected a Vintage Square One wave in each and routed them to bus one (so that their full outputs go through the filter). You will see that oscillator six has been raised by an octave.

Adding these two oscillators gives the patch is wonderful tone with the nuances still being controlled by the simple FM stack.

The final tweak I made was to add some effects. In this case, I added a Quad Phaser and some Plate Reverb. The phaser slightly thins the tone giving a bit of movement and the reverb adds some space to the sound.

I know I'm biased, but I love this patch. I think that by coupling FM and subtractive synthesis you can create really playable sound that uses the best elements of FM and subtractive without having the disadvantages of sounds created using only one technique.

Z3TA+ Piano Developed

I then developed Z3TA+ Piano by making two changes:

- adding oscillator four to act as a modulator for oscillator five, and

- adding LFO five and LFO six to modulate the filter and oscillator three's volume respectively.

When listening to the patch on its own (in other words, not within the context of a track) I prefer Z3TA+ Piano to Z3TA+ Piano Developed—you may have a different view. To my ear the former is more playable and this latter patch is too bright. However, within a track this developed patch may work better. Anyway, let me explain what I did in a bit more detail.

Oscillator four is loaded with a sine wave that has its level set to 62.5%. This frequency modulates oscillator five. The output level of oscillator four is modulated through the modulation matrix by envelope three with a Range of 0 to 34% and, as with the modulation of oscillator one's Level, the modulation of oscillator four's Level is controlled by velocity.

Envelope three (which acts as the modulation source for oscillator four's output Level) is set up as an ADSR type envelope. The Attack Time is set to zero and the Decay Time to 0.03 seconds with the Sustain Level at zero. The effect of oscillator four is to give more bite in the attack and to make the tone thinner. It has no effect on the sustain portion of the note.

The other main change is the addition of two LFOs. These add a slight wobble to the tone and an element of tremolo. For these elements I used LFO five and LFO six which are both polyphonic LFOs and each has its phase set to free. This means that their effect on each separate note will be slightly different.

LFO five was set as the modulation Source for the filter Cut-Off. The Speed was set to 9.4 Hz and the Fade time to 0.28 seconds. In the modulation matrix, the Range was set to 21.9% and the U-Lin- curve selected so that as the LFO wobbles it reduces the cut-off frequency. As the note sustains, this addition gives a slight wobble to the tone and a very slight tremolo effect.

However, although a polyphonic LFO has been selected, the tremolo is rather uniform. To keep the effect but remove the uniformity, LFO six modulates the level of oscillator three. The Speed of this LFO was set to 2.21 Hz, the Fade to 1.31 seconds, the Range to 43.8% and again the U-Lin- curve was used.

The combination of the two LFOs gives a far more natural tremolo type effect.

Lead

Lead sounds are intended to feature prominently in a mix without dominating.

Z3TA+ Lead

I am going to demonstrate one lead sound using Z3TA+.

Lead Gtr

This patch makes extensive use of Z3TA+'s effects units, however, the other elements of the sound are still important.

The sound is intended to play lead guitar-type parts and single note guitar-type riffs. While this patch is intended for single notes, the polyphony has been set to four so that sustained notes are not unnaturally cut off. Like a real guitar, this patch is highly touch responsive—at lower velocities, the tone is muted (although clearly very overdriven) while at higher velocities the sound screams.

The root of this patch is fairly simple: two sawtooth waves (one two octaves above the other) which are routed into a 36 dB filter. This gives a slightly distorted tone and nothing more. The filter Cut-Off is then closed down to 48 Hz and the Resonance increased to 8.17 dB. At this point no sound is audible. To get some sound and to give some shape to that sound, the filter cut-off was modulated twice in the modulation matrix:

- first by envelope one—here the envelope adopts a fairly conventional ADSR envelope with a Decay Time of 2.41 seconds and Sustain Level of 70%, and

- second by the key position so that the envelope opens up more in the higher key ranges. In the modulation matrix, the source is On—the Range 7% to 70%, the Curve Slow+ and the control is U-Note#.

Finally some vibrato was added (which can be controlled by the modulation wheel). The modulation Source for the vibrato is LFO One which uses the Triangle wave and had its Speed set to 7.68 Hz. The Range is 35% and the Curve is Pitch 1S (one semitone, if the full range is selected).

For this patch, I set the vibrato separately for oscillator one and oscillator two. It would have been possible to select All Osc Pitch as the modulation destination. However, that would rob the patch of the flexibility when adding other oscillators. In addition, this architecture also means that different ranges can be selected (which can dirty up the vibrato).

At this point you can hear that there is a broad spectrum of tone ranging from almost a harmonic like sound at low velocities to quite a gentle tone in the mid-range through to a more biting tone at the higher velocity ranges.

However, we don't really have the scream of a lead guitar, so this is where we call on the FX units.

There are three key elements that take the basic sound and give it the characteristic of the finished sound—the Heavy Metal distortion unit, the Six-Voice Chorus and the Mid Wide EQ.

The Heavy Metal distortion was chosen for its tone. The Gain was set fairly high at 12 dB, but not as high as it could be, and the Level was set at 71%. To control the tone, the output was routed to filter one: without this routing, the tone is just thin and buzzy and really unusable. To finish of this stage, the Tone was set at 78% and the Decimator to 7%.

After going though the distortion unit, the sound needed to have some of the thickness taken out and to be made brighter. For this the Mid Wide EQ was called upon—the 1.2k band was dropped by 5 dB, the 3.2k band was boosted by 12 dB and the 5k band was boosted by 2.25 dB.

However, even with the distortion and the EQ, the key to taking the distorted tone and making is thick and smooth is the Six-Voice Chorus which also has the effect of spreading the patch over the stereo field making the patch very dominant over a wide area of the sound spectrum. With the Six-Voice Chorus, the Depth was set to 62%, Speed to 0.09 Hz, Delay to 9ms, Feedback to 30%, the EQ Mode was Hi+ and the Low control was set to -12 dB with the High to 0.63 dB and finally the Level was set to 60%.

You will hear that even with these effects being added—and some are quite extreme—the patch is still very velocity responsive.

Once the distorted lead guitar tone had been set it was then embellished with a Ping Delay and a Plate Reverb. Both of these were added to taste, in particular the echo was set to fit a riff for which this sound was programmed.

Pads

First a word of caution. We're going to create some pads here—many of these are very thick and will fill up huge amounts of the sonic range.

In my personal opinion, one of the key differences between professionally and non-professionally produced tracks is the use of pads: professionals tend not to use them (or if they do, they are used with subtlety). By contrast, the amateur producer will often use a pad as a first step to fill out a track.

Feel free to use pads in your productions, but please, think how you are going to use the pad and don't use it to cover other shortcomings in your production.

Rhino Pad

The first pad I am going to create is with Rhino.

FM Pad

We'll create some incredibly thick pads later with Z3TA+. I wanted FM Pad to be a more delicate pad which could therefore be used in more situations.

This pad is essentially two simple FM stacks where the level of the modulator fluctuates to give subtle tone shifts over time—we will use a similar technique in Male Voice Choir (see below) but with this latter patch the tremolo will be introduced by an LFO. In this patch I won't try to describe the volume envelopes in detail: you'll have to get the patches if you want to take a look.

For the first stack, operator one is modulated by operator two with a 1:1 ratio—in the routing matrix, the modulation amount was set to 47. In the second stack, operator three is modulated by operator four with a 2:1 ratio and in the routing matrix, the modulation amount was set to 34. This second stack was slightly detuned (both operators by +9).

To give a slight bit of extra tone to the first stack, operator five modulates operator one with a 1:6 ratio—the modulation amount was set to 3. To give this tone a bit more movement, operator five is modulated by operator six which acts as a low frequency oscillator.

The outputs from operators one and three are then fed into both the raw output and filter one—the respective levels are balanced by the Filtered slider. You will see that I have set this slider at 70. The filter selected is Analog Lowpass 3 and the Cut-Off slider is set to 4.

As you can hear, the result is a simple, but effective pad with the ability to shift the tone by moving the Filtered slider.

Vanguard Pad

The next pad is created with Vanguard.

Thick Pad

With Thick Pad, I wanted to create a simple, but quite dense pad. The end result gives a tone similar to Lush Pad (which we will look at later) but uses considerably less system resource.

This patch uses three waves, a Sawtooth, PWM Saw and a Square wave—the PWM Saw was tuned an octave higher than the other two, the Fat knob was turned to 96, and VCS (voices) set to 7. This combination gives a very thick, bright sound.

To take this from a bright patch to a pad, the 12 dB low-pass filter was called up with the Cut-Off set to 37 Hz, the Resonance to 20% and the Keytrk to 55%. This makes the sound dull. To give some life, envelope one was called up and the Cut-Off knob (in the envelope lane) turned to 22. The envelope was then set with an Attack time of 2.2 seconds, a Decay time of 4 seconds, a Sustain level of 104 and a Release time of 2.9 seconds.

To give the patch volume some shape, the Level knob in the envelope one lane was set to 127. In the Amp section, the Volume was set to 40, the Veltrk to 32 making the patch quite touch responsive, and the Spread knob was set to 80 with the Drive knob to 28 to add a bit of dirt.

To give the sound a bit more movement, oscillator one (Sawtooth) and oscillator two (PWM Saw) are modulated by their LFOs. LFO one was set to 8.96 Hz with the Detune knob to 19. LFO two was set to 3.44 Hz with the Detune knob to -14 and the PWM knob to 44.

I could have called up some FX for this pad, however, I didn't feel that they would have added much and also, they would have used more system resource.

Z3TA+ Pads

This next group of pads is created with Z3TA+.

Male Voice Choir

I wanted to create a patch that was evocative of the sound of a male voice choir. It took me quite a while to get the precise sound I was after. For those of you with the patches I have included two patches which, although I really like the sound, I rejected because they were not the precise sound I was searching for.

The two patches I rejected, Voices and Voices II, are both based on a square wave pushed first through a 24 dB low-pass filter and then through a formant filter. Voices II has more layers, more vibrato, and a higher pitched ghost-like element.

The two patches have quite an ethereal, haunting quality to them. However, although they are breathy and might in certain circumstances sound like an effected human voice, they are not close enough to the male voice choir sound I wanted.

With MVC Root, I found the essence of the sound I was looking for. Again, a square wave is pushed through a format filter. However this time the square wave is frequency modulated by a sine wave. This element of frequency modulation changes the tone radically and gives it that "male voice" quality that I was looking for.

If you listen to MVC Root you may not be that impressed with it—especially after Voices and Voices II which are both much more developed sounds. You may also disagree with my assessment that the sound is reminiscent of a male voice. Push the modulation wheel and play the note and you will hear a radically different quality: the amount of FM is controlled by the modulation wheel.

I suggest you move the modulation wheel before you play the note—if you move it too quickly while the note is sounding the effect is not particularly pleasant. Much like a formant filter, there are a few quite distinct tones to be found in the travel of the modulation wheel—shifting too obviously from one tone to another does not necessarily give a pleasing result. If you want to add a bit more life to the sound, you can use your expression pedal to control the vibrato.

In this patch I use FM to create the root of the sound. To create the source sound I could alternatively have used the Waveshaping tool or imported my own wave. However, I chose FM first because it can be controlled over time (which the other two options cannot) and more importantly because I'm happier using FM.

Anyway... that was how I got to the root for Male Voice Choir. This is how I worked from that point to create a patch. When auditioning this patch, the range when an acceptable male choir sound can be obtained is quite narrow—if you play outside of this range, the effect may be lost. This reflects some of the limitations of the human voice.

Oscillators one and two create a simple FM stack with a C:M ratio of 1:1. This whole stack is doubled by oscillators three and four which create another simple FM stack which uses a C:M ratio of 1:2. In both cases, the modulator's wave is a Sine wave and the carrier wave is a Vintage Square Three wave. Both of the stacks then feed directly into the filter block.

In the first FM stack, the Level of oscillator one was set at 10.9%. However, this level is then modulated by LFO one (which was set to a Triangle wave with a Speed of 3.85 Hz and the Range was set to 11% in the modulation matrix). This subtle modulation causes the level of oscillator one to fluctuate gently. This variation in the volume of oscillator one causes the tone of oscillator two (which is frequency modulated by oscillator one) to waver slightly. The sonic result is that the output from oscillator two takes on some of the inconsistency of the human voice. In other words, this modulation of a the modulator allows us to achieve precisely the result we are looking for.

In the second FM stack, oscillator three's Level is set at 30% and is modulated by LFO Two (which was set to a Triangle wave with a Speed of 2 Hz, and the Range was set to 30% in the modulation matrix). Again this modulation gives the inconsistency to mimic the human voice.

With the two FM stacks (created from the first four oscillators) working together, oscillator two was detuned by 12.5 cents to give a bit more movement to the tone and oscillator two and four had their Levels set at 64% and 54%, respectively.

To give some further movement to the sound, the pitch of oscillators two and four are then each modulated by a Triangle wave LFO. Oscillator two is modulated by LFO Five with a Speed of 1.14 Hz and a Range of 20%, using the Pitch 1S curve, and LFO Six modulates oscillator four with a Speed of 2.96 Hz and a Range of 19% using the same semitone curve. The polyphonic LFOs have been engaged so that each note is separately modulated giving more of a choral effect.

The filter that is used is comparatively simple—there is no modulation. The filter for the FM elements (filter one) uses the Formant filter with its Cut-Off frequency set to 53 Hz and its Resonance set to 0 dB. This cut-off frequency gives an "A" (as in cAt) tone. The second filter is used later and does not affect this element of the tone.

Before I continue with the construction of this patch, let me meander for a moment and explain a bit further (with an audio example) why I used FM for this patch. As you will have noticed, the FM effect we introduced is very subtle. Indeed, I would not blame you for being skeptical about its effect. If you are skeptical, then turn off oscillators one and three and listen to the results. Without these two FM modulators, the sound takes on a totally different tone becoming far more metallic and harsh. There is absolutely no way that the sound without the FM could be described as having a vocal quality, even with the formant filter.

Returning to the patch, we need to give it a bit of shape—this is where we tweak two of the envelopes. First, to set the volume of the whole patch, the Amplifier envelope was edited. A fairly simple ADSR envelope was set up with the Attack Time set to zero, the Decay Time set to 0.51 seconds, the Sustain Level set to 68% and the Release Time set to 2.9 seconds.

These settings were largely chosen so that they didn't have too much effect on the settings chosen for the envelopes which modulate the oscillators—in particular, the release time was set so that it didn't cut the release time of the oscillator envelopes. It would have been possible to use the amplifier envelope to control the output of oscillators two and four. However, I didn't particularly like that effect, so I used the separate envelope. Equally, I felt it was important for this patch to use the amplifier envelope to give a bit more shape to the sound.

For the outputs of oscillators two and four, I set envelope two as the modulation source using the full range in the modulation matrix (so the levels of oscillators two and four are fully controlled by envelope two up to each oscillator's output level). Envelope two was set as an ADSR type envelope with the Attack Time set to 1.05 seconds, the Decay Time set to 4.4 seconds, the Sustain Level set to 68%, and the Release Time set to 1.18 seconds. This gave a feel reminiscent of a choir.

At this point, the sound of the patch was interesting but I wanted to add some thickness, particularly in the lower registers. To do this I called up oscillator five and loaded a Vintage Saw One wave in Multi Free mode (multi-mode with the phase of each oscillator running freely). The Phase slider was set to 104 degrees to give a fairly fat sound.

Simply adding an oscillator didn't do enough—this gives a fizzy sound which detracted from the choir sound created by the first four oscillators and filter one. To remedy this I routed the new oscillator through filter two which is a 36 dB low-pass filter with its Cut-Off set to 675 Hz, and its Resonance set to 3.4 dB. This routing means there are now two separate elements to the patch which are layered together.

While the filtering of this fifth oscillator improves the sound, it still didn't quite get there in terms of fitting with the rest of the patch. To help it fit, I did two things:

- First, I added some key scaling to the output level of filter two. This has the effect of reducing the volume of this fifth oscillator/second filter combination as higher notes are played—this gives the lower range thickness while keeping the higher range tone. To do this I set B-Note# as the Source in the modulation matrix with Filter Two Level as the Destination: the Range was set to maximum and the curve to U-Lin-. It is this negative curve that leads to the damping of the volume at higher pitches.

- Second, I added a volume envelope to modulate the output of oscillator five. In the patch this is envelope generator one and in the modulation matrix, the Range is full and the Control is Velocity. The envelope is very simple: the Attack Time was set to 1.3 seconds (so it is slightly slower than for the other two sounding oscillators) and the Sustain level was set to the maximum level (hence the Decay Time has no effect). The Release Time was set to 3.16 seconds, and so sounds for slightly longer than the "vocal" oscillators after the note is released.

There is only one velocity sensitive feature in this patch: oscillator five's output. The velocity control means that with harder playing the patch sounds thicker in the lower regions. Given the nature of the patch, I didn't want there to be much in the way of velocity modulation.

The effects are comparatively subtle for this patch. First there is the Six-Voice Chorus which adds smoothness, brightness, and stereo breadth to the sound. Secondly, there is some Plate Reverb which adds some smoothness and space.

Lush Pad

Here's a pad that will melt your CPU. It is also a very dominant pad and so you may not be able to add it to every track you create. This is a comparatively simple pad, but it does use the effects units fairly heavily. However, the effects are used to enhance the tone rather than to radically re-shape the sound.

The basis of the pad is three oscillators, each with the Vintage Saw One wave selected. For each of the waves, multi-mode (Multi, Free) has been selected and the Phase set to 90 degrees, so from the start we're using a fair amount of CPU. One oscillator is split between the two busses, one turned predominantly to the first bus and the last oscillator turned predominantly to the second bus. There are two, parallel 36 dB low-pass filters: each with their Cut-Off set to zero and the Resonance set to 2.59 dB. One filter is then panned left and the other right.

To complete this basic set up—as there is no sound with the filters being closed—envelope one was set as the modulation Source modulating all of the filters' Cut-Off frequencies. The Range was set to a Minimum of 21% and a Maximum of 89%. Finally Velocity was set as the Control. Envelope one was set up as a simple ADSR envelope with Attack Time set to 0.04 seconds, Decay Time to 0.51 seconds, the Sustain Level to 87%, and the Release Time to 3.45 seconds. In addition, in the Amplifier envelope the Attack Time was set to 1.46 seconds and the Release Time to 3.45 seconds.

This gives the main sound to the patch—to give some more thickness, and change the tone a bit, a fourth oscillator, Vintage Square Two, was added. This time a single wave, not a multi-mode wave, was called up which is tuned to the same pitch as the sawtooth waves and balanced between the busses.

I only wanted this fourth oscillator to thicken up the tone a bit in the lower regions so I added some key scaling to reduce the level at higher pitches. To do this, in the modulation matrix I set the Source to B-Note#, the Range to full, the Curve to B-Lin- and oscillator four's Level as the destination.

This is the main sound. To give it a bit more richness and movement and to emphasize some of the inherent characteristics of the patch, in particular some of the velocity nuances, I added some effects.

First, to keep some of the higher frequency elements in check (so keeping the pad thick) I selected the Soft Drive distortion unit. Here I added some Drive (7 dB) and some Level (23%), kept the

Tone full and routed the output back to the filters. Next I added the Six-Voice Chorus to give some more swirl to the sound and finally I added a Ping Delay and some Plate Reverb.

Lush Pad Developed

I like Lush Pad, but I also wanted to take it in a slightly different direction, hence Lush Pad Developed which makes three changes to the basic theme:

- The first change I made was to arrange the filters in series and pan them to the center.

- Secondly, I swapped the second filter for a 12 dB high-pass filter.

- Lastly I set the low-pass filter to open gently during the release stage giving the impression of a lighter sound and longer sustain.

With the changes in the filter block I switched the modulation matrix Destination so envelope one only controls the Cut-Off of filter one. I then introduced envelope two as a modulation Source for the Cut-Off of filter two.

Filter two's Cut-Off was kept at its minimum but the Curve for envelope two was set to U-Lin-. This means that the envelope works to keep the cut-off low (so that the sound can be heard). Envelope two had its Attack Time set to 2.4 seconds—you will hear that the sound fades in: this is the high-pass filter having its cut-off frequency swept from high to low giving the fade effect. The Sustain Level was set to 100% so the Decay Time has no effect. Finally the Release Time was set to 5.2 seconds so that the sound gets progressively thinner during the release phase.

To open the low-pass filter during the release stage, thereby making the sound progressively brighter as it decays, I called up the Pitch Envelope. The Pitch Envelope does not have to be used only for pitch—in this instance I took advantage of its bipolar behavior. It has no effect until the release stage when it comes into action as a modulation source with a Release Time of 2 seconds and a Release Level of 100%.

The Pitch Envelope acts as a modulation Source to filter one's Cut-Off through the modulation matrix where the Range was set to 100%. Unlike all of the other envelopes the Pitch Envelope's release level can increase during the release phase so the modulation routing in this instance is quite simple.

Pluck and Stabs

As a general rule, plucks and stabs will often be prominent, but only momentarily.

Rhino Plucks

This first pluck-type patch is created with Rhino.

Guitaresque

Those of you with the patches will see that I have included a patch called Guitar-Unesque: this is my failed patch. For Guitaresque I wanted to create something close to the tone of a clean guitar played through a chorus pedal—ideally I was looking for the kind of tone that could be used to play guitar-like arpeggios. You will understand that Guitar-Unesque didn't go in this direction so I gave up.

Guitaresque has two main elements—a thin, almost sitar like sound and another sound which in certain regions sounds rather like tuned wooden percussion. In combination, we get a wholly new sound which is much closer to a clean guitar sound.

The first part is created by a parallel modulator block. Operator one is modulated by operator two (the modulation amount is 72 with a C:M ratio of 1:3) and operator three (the modulation amount is 74 with a C:M ratio of 2:1). In addition, operator three also modulates operator two with a C:M ratio of 7:1, here the modulation amount is 68.

All of the envelopes in this patch decay swiftly with a curve set to about 25. For operator one, the Level envelope decays to zero after about six seconds. For operator two, the Level envelope decays to zero after about 0.8 seconds and for operator three, the envelope decays to approximately 50% (where it sustains) after about 1.2 seconds.

For all three operators there is some velocity scaling. The scaling in operator one ranges from approximately 20 to 100, in operator two from approximately 25 to 100 and for operator three from approximately 50 to 100.

The second part of the patch comprises two elements. First operator four is routed through the Filter and to the Raw output (both set to 100). The waveform selected for this operator is the Sine wave tuned an octave below operator one (although not modulating that operator). The envelope for operator four decays to zero over about six seconds.

The prime purpose of the operator is to thicken up the sound—routing it through the filter doesn't change its tone much, but it does double its volume.

The last part to this sound is operator six which is modulated by operator five with a C:M ratio of 1:15 and the modulation level set to 68. The envelope for operator six decays to 15% over approximately 1.2 seconds whereas the envelope for the modulator, operator five, decays to zero over approximately 0.8 seconds (although this operator uses a much steeper curve). This last combination gives a the tuned wooden percussion type sound.

This final simple FM stack is then routed through Filter One where Analog Lowpass 1 has been selected. The cut-off slider was set to 28 and the envelope slider (Cutoff Env Mod) was set to 55 and the cut-off envelope decays to zero over approximately 6 seconds.

To round off the patch, the Stereo Ensemble and 8 Taps Reverb were called up in the FX section.

It would have been possible to use a pluck type wave to get plectrum sound, however, I'm not a big fan of this effect. First, I find it hard to control the waves in Rhino to give a plausible effect and second I find that the picking sound can then dominate a patch.

Vanguard Pluck

This next pluck patch is created with Vanguard.

217

Thin

With Thin I wanted, quite literally, a very thin sound that would work as a plucky sound but wouldn't dominate the mix too much. I wanted a sort of clavinet type tone, but more subdued and with far less bite.

The sound source for this patch is Combed Square Two and Combed Square Three: these are quite bright waves with almost a metallic quality. To simply tame this sound with a low-pass filter would not work with this patch—it would only make the sound dull rather than thin, so for this patch I chose the filter—M—which is a bandpass filter. The Cut-Off was set to 2 kHz, the Resonance to 32% and Veltrk to 100%.

This filter setting gave quite a dull sound, so envelope two was engaged to modulate the filter cut-off. The Attack time was set to zero, the Decay time was set to 91ms, the Sustain level was set to 7 (on a scale to 127), the Release time was set to 142ms and the Cut-Off amount in the envelope two lane (the amount by which the envelope modulates the cut-off frequency) was set to 58.

As a last step, to give the patch a bit more shape, envelope one was called up to modulate the Level. The Attack time was set to 0ms, the Decay time to 57ms, the Sustain level to 62, the Release time to 104ms and the level knob to 114.

Wusikstation Stab

Wusikstation is also very capable of creating fat stabs.

Saw Stab

With this patch, I wanted to create a big multi-mode type oscillator patch in Wusikstation. There are several reason why you might want to do this. First Wusikstation generally uses less CPU than many synthesizers (this is especially so as Wusikstation has the same CPU hit whether it is using a sample of a single wave or a multi-mode wave) and the second reason for this patch is that I like the sound.

For this patch, the Super Saw wave from the Famous Keys Volume Two bank was loaded into the first three layers. These are panned left, center, and right—one is kept at its pitch, one fine tuned up by 17 and the other fine tuned down by 23.

Each layer is fed through a low-pass 4-pole (24 dB filter) with the Cut-Off set to zero—in layer one, a small amount of Resonance (11) was added. The filter cut-off for all layers is controlled by Modulation Envelope One through the modulation matrix: the Source and Destination, should be obvious, the Minimum and Maximum were set to 0 and 127 respectively, and the Amount for layer one was set to 108, for layer two 99, and for layer three 96.

Modulation Envelope One which controls the three layer filters is a simple affair: the Attack and Decay time were set to zero, and the Sustain level set to 127. The Vel knob was set to 114 which gives a wide range of control over the cut-off of the filters.

Layer one has some resonance set in the filter. I wanted to add some resonance for the filters in layers two and three. However, I wanted the resonance to be greater at lower velocities and less at higher velocities. To do this I called up Modulation Envelope Two and set it very much like Modulation Envelope One but with the Vel knob set to 43. I then set the envelope to modulate the Resonance in layers two and three—to increase the resonance at lower levels, I set the Minimums to 127 and the Maximums to 0. The Amounts were then set to 51 (layer two) and 49 (layer three).

As a final tweak Modulation LFO One was called up to modulate the Fine Pitch of layer one (which was the layer without any detuning). The LFO wave is a Sine wave and the Speed was set to 1/32. In the modulation matrix, the Minimum and Maximum were set to 0 and 127 and the Amount to 58.

The patch was finished off with some effects. All of the layers were sent to FX1 (127)—in this slot the Stereo Echo was selected with a Delay Time set to 1/12. No layer was sent to FX2, however, the full output of the echo was fed into the Lite Verb reverb unit which was called up in FX2. The FX were structured in this way so that they add weight to the patch but are kept further back in the sound field.

Z3TA+ Plucking and Stabs

As you would expect, Z3TA+ is very capable when it comes to creating plucks and stabs.

Bad Guitar

This is a very simple patch which has a tone almost reminiscent of a bass guitar being slapped in its higher registers. As a patch it could be developed further, but I like the simplicity and so didn't want to change it. However, please don't let me stop you.

The sound source for this patch comprises two waves: in oscillator one the Fret wave has been selected and in oscillator two, a Sine wave has been selected. The fret wave gives the patch its tone and the sine wave gives the warmth and thickness to the sound without obviously being present.

The waves then run into the filters which are arranged in series (by clicking on the Dual button). Filter one is a 36 dB low-pass filter with the Cut-Off set to 16 Hz (fully closed) and the Resonance set to 0 dB.

To open up the filter, envelope one was set as the modulation Source in the modulation matrix with filter one's Cut-Off frequency as the modulation Destination. The Range was set to a Minimum of 42% and a Maximum of 100% with Velocity being set as the Control.

In envelope one, the Attack Time was set to zero. To mimic the behavior of the envelope when a string is picked, the Slope Time was set to 0.34 seconds and the Slope Level to 54%. To create the decay section of this patch (which does not mimic the decay of a real instrument) the Decay Time was set to 1.32 seconds and the Sustain Level to 32%. A bit of shape was also added in the Amplifier envelope.

The second filter is a 12 dB high-pass filter—its function is to remove all of the muddy elements from the patch. The Cut-Off was set at 180 Hz and the Resonance at 1 dB (which gives a subtle emphasis to the fundamental tone).

The effect of the two filters is to create a great tone, however, they have also taken a lot of energy out of the patch which is remedied in the in the effects section. The tool for this job is the compressor—which also has the effect of evening out some of the harmonics giving a brighter but more rounded tone.

The use of the compressor in this patch lacks any subtlety, however, I like the result. The Fast mode was selected, the Threshold set to 0 dB and the Ratio to 50:1 (in other words, the compressor is acting as a limiter). The Gain was then pushed up to 12 dB and the Level set at 100%.

Finally the Six-Voice Chorus was added to give some brightness and sparkle. To ensure the chorus added brightness, the Hi+ EQ was selected and the Hi control pushed up to 4 dB.

Muted Stab

For this pad, I was trying to create a muted stab with a lot of tonal variation. For me, it was important that this pad be muted but still have character.

I did initially start to build this patch in Vanguard—listen to Unwanted Stab. However I abandoned that patch because I didn't like the character and it was too close to Gentle Stab. One factor to note with Z3TA+ is that its patches tend to have quite a dominant character that cut through mixes—you can hear Muted Stab clearly in a mix even though it is a comparatively subdued patch.

The basis of the patch is two slightly detuned Vintage Saw Four waves each with multi-mode (free) engaged (each with the phase slider pushed up to 90 degrees) and panned in opposite directions to about 70/80%. These are then fed into two 36 dB low-pass filters in series: I'm not certain how often you're going to need a 72 dB/octave filter (as is created here). However, for the tone I was after at lower velocities, this combination worked for me. You should, of course, be aware that this combination will use more CPU than a single filter with a gentle filter slope.

In both filters, the Cut-Off was set at 16 Hz (so no sound passes through) and the Resonance was set to 4.94 dB. To give the filters some movement, envelope one was set at the modulation source with all filters Cut-Off as the destination. The Range was set to 83% in the modulation matrix and the Control is Velocity. Finally, I selected the U-Lin+ curve.

The effect of the U-Lin+ curve is subtle—to my mind its effect makes the control of the filter easier. At higher velocities there is little difference, however, at lower velocities it is easier to get (and control) that "muted" tone that I am after.

The envelopes were set as simple ADSR envelopes. Envelope one which controls the filter was set with an Attack Time of zero, a Decay Time of 0.21 seconds, a Sustain Level of 60% and a Release Time of 0.33 seconds. The Amplifier envelope was also set with an Attack Time of zero and then a Decay Time of 0.06 seconds, a Sustain Level of 75% and a Release Time of 0.33 seconds

To then give the patch some more width, I panned the first filter mostly to the left and the second filter mostly to the right. This also has the effect of changing the tone of the patch slightly—it is still muted, but probably brighter/warmer with the panning.

Finally to round off the patch I added some Plate Reverb.

Separated Stab

On first hearing, Separated Stab sounds similar to Muted Stab which is not surprising being as I copied the earlier patch and tweaked it. However, from a playability perspective, the patches are quite different. The components of the patch are also different.

With Separated Stab, I was looking for a brighter tone that would be more dominant. However, I also wanted to have more control over this patch.

The setup for Separated Stab is similar to Muted Stab, the key differences are that the Octaved Square One waves are used (instead of Vintage Saw One waves) and the filters are run in parallel (not in series).

Envelope one uses the same settings as in Muted Stab although there is a change to its deployment through the modulation matrix with the Range being set to a Minimum of 29% and a Maximum of 84% (and remember that the filters work in parallel, not in series for this patch).

The tone of this patch was good, but it lacked some of the thickness of Muted Stab. To rectify this I added a third oscillator using the Vintage Saw One wave. This gave a thicker tone, but to my mind the effect was more pleasing when the pitch of this new oscillator was raised by two octaves. The third oscillator also robbed the patch of some of its subtlety. To remedy this, and control the operation of oscillator three, I turned to the modulation matrix.

In the modulation matrix, I added a line with the Source set to On, the Range to Maximum, the Curve to Slow+, the Control to Velocity and the Destination to oscillator three. This means that oscillator three has less effect at lower velocities (so the sound is less fat) but more effect at higher velocities (making the patch thicker).

To give some more tonal variation to the patch—and to reduce some of the power of the resonance—I added another line to the modulation matrix. The Source was set to On, the Range to 52%, the Control to Modulation Wheel and the Destination was All Filters Separation. By separating the filters more tonal changes can be uncovered without robbing the sound of its essential tone and volume. This change can be controlled in real time with the modulation wheel or automated.

The sound was still a bit thick too my ears. As there were no filters available, I turned to the EQ section and selected the Wide One EQ mode and dropped the 80 Hz slider by -18 dB. The effect of this tweak is to take out some of the low end from the patch leaving more room for the bass and the kick drum in the mix.

Finally I added some Plate Reverb to taste.

Chapter 12—Refill #1

Arpeggiators, Step Generators, and Other Playback Tools

In the original free version of this book which I published in 2005, I said that there would be two refills published in the latter half of 2005. OK. I'm sorry. I'm kinda late with the refills. However, I did say you would get two refills. Instead, you're getting four refills, plus a new chapter supplement, *and* the whole book has been revised and expanded, so I won't feel too bad. Added to that, the book is now available in hard copy format from all leading bookstores: if you're reading the free refill, go and buy the book on Amazon!!

One thing I did say in chapter 14 was that the first refill would look at arpeggiators and other playback tools. You know me... I'm always keen to keep my word, even if I'm a bit late, and so this first refill looks at arpeggiators, step generators, and other playback tools.

The key focus here is on tools that will create melody and/or rhythmic effects without the need for external automation. The benefit of this sort of tool is that you can create rhythmic and melodic programming which is both compelling (from a sonic perspective), and which cannot be readily created in any other manner.

Most of the techniques used in this refill have already been described in the book. However, this refill is a good opportunity to bring all of these tools together, and to demonstrate some of the benefits of these techniques.

Playback Tools in Z3TA+

Z3TA+ has a range of tools which allow us to animate our playback: some tools are visible on the face of the synthesizer, and others are hidden.

12 Step

12 Step demonstrates one of the most obvious tools for adding some rhythm to a sustained note or chord, the LFO.

This patch is based around a single Vintage Saw 1 wave with Multi, Free mode selected and the Phase slider pushed to 125 degrees. This then feeds into a 24 dB low-pass filter with the Cut-Off set to 16 Hz and the Resonance to 1.36 dB. The cut-off is then modulated with an envelope (EG1) with the following settings:

- Attack Time: 0 seconds

- Slope Time: 0 seconds

- Slope Level: 100%

- Decay Time: 0.45 seconds

- Decay curve: exponential (convex)

- Sustain Level: 56%

- Amount: 100%

The effect of this envelope is to give a slight emphasis to the note when it is first struck, but apart from that, the only effect is how the Sustain Level keeps the filter slightly open.

Many of the Z3TA+ patches in this section use this same basic foundation.

Through the book we've already looked at patches where an LFO works as a modulation source, controlling volume to give a stuttering type effect. To an extent this patch is no different. However, one feature that differentiates Z3TA+ from much of its competition is its wide range of LFO waveforms. These allow for different effect to be created.

This patch uses the 12 Step LFO waveform with the LFO's speed set to 7.16 Hz. The sonic effect of this waveform is to give a constantly shifting volume rather than a simple on/off type rhythm. This is probably a far more subtle sound than can be achieved with most synthesizers.

Load the Gate File

This second patch is called Load the Gate File. Why? To remind you to load the gate file. If you have purchased the patches, you will also find a MIDI file called Z3TA Gate.mid. Open up an instance of Z3TA+ in your favorite sequencer, load Load the Gate File in Z3TA+, open up the MIDI file, and set the MIDI file to play in a loop. Then start to play some chords.

The MIDI file creates a rhythmic pattern playing C0 notes. When the C0 note is played, the audio output of the synthesizer is closed down (so no sound gets out). This gives the characteristic rhythmic sound that you can hear.

You don't need a MIDI file to use this feature: just hit the C0 key and the audio output is muted—I've only included a MIDI file to demonstrate this feature.

Arpegi8r

I don't think you need this book to tell you how to use an arpegiattor. This patch is quite straight-forward and is based on the multimode sawtooth waves that were used in the two earlier patches (with the same envelope settings). The arpeggiator (in the LFO section) was then set as follows:

- Pattern: Up

- Sync: 1/4

- Octaves: 2

- Length: 57%

There are two interesting features of this arpeggiator pattern which are useful in creating compelling rhythmic sounds:

- First, all of the envelopes are retriggered by the arpeggiator. This means that each time the arpeggiator generates a new note, the envelopes restart.

- Second, the Length control allows you to control the length of the note played by the arpeggiator. This allows you to play a shorter note and thereby create a more staccato effect, but without messing with the envelope controls which may lead to a different tone being created.

By the way, if you don't know what an arpeggiator is, and you've never played with one before, load up this patch and hold down a chord. Instead of hearing a group of notes playing together, Z3TA+ will play each note of the chord (and the note one octave above them) in turn as an arpeggio. Since this patch has the Up pattern selected, the notes will be played in ascending order.

MIDI File Player

One feature of Z3TA+'s arpeggiator is the facility to play MIDI files. This can give stunning results: listen to the default Z3TA+ sound, 8-Bar First Contact, for an example of this in practice. The downside to this feature is that when you try to play a chord, you are likely to get a very messy sound.

MIDI File Player creates its sound by playing a MIDI file. For your interest, the MIDI file (called MIDI File) is also included with the patches.

With this feature you can load any single bar MIDI pattern and create a rhythmic pattern (obviously the rhythm depends on the source MIDI file), which can then be readily transposed. The MIDI file can also include other MIDI data (for instance, you could include a filter sweep).

To load the MIDI file, all you have to do is select Options > Import MIDI File to Arpeggiator. If a MIDI file is loaded, its name will show next to this menu item. Once the menu item is selected, in the arpeggiator, set the Pattern to MIDI File and set the Sync or the Speed. Just because the imported MIDI pattern is restricted to one bar, that doesn't mean the Sync has to be a division of one bar—patterns can last longer than a bar by setting an appropriate Sync length.

By the way, if you do have the patches, you may encounter a minor quirk with Z3TA+ leading to you not hearing anything, or to the patch not playing properly. If this happens, go to the arpeggiator in the LFO section (click on the AR button), and against Pattern, it should show MIDI File. Right-click and then left-click on the text that says MIDI File and you should see the MIDI File text again. However, you will now find that the pattern plays. If you're still having problems, try reloading the MIDI File (and then right-clicking, left-clicking again).

Playback Tools in Vanguard

Vanguard gives us two main playback tools:

- **Trancegate.** The Trancegate is a simple, but incredibly effective device which has spawned many imitators, that allows for the creation of rhythmic patterns. It works by breaking the audio signal into a number of steps synchronized to the host's tempo. For each step, the audio either plays or doesn't, thereby creating a rhythmic pattern. The Trancegate in Vanguard can work on a stereo basis allowing each channel to be independently gated so the signal can be readily thrown about in the stereo spectrum.

- **Arpeggiator.** The arpeggiator in Vanguard works very much like an arpeggiator in any synthesizer (although it doesn't have the MIDI file playback features that Z3TA+ offers).

I'm going to create two patches in Vanguard which demonstrate these two features. Both of there patches are based on a Sawtooth wave with a Square wave pitched an octave lower. The VCS (voices) control is set at 7 and the Fat control to 54.

T-Gate

T-Gate takes advantage of the Trancegate. The Speed has been set to 1/16 and the Contour to 127 so that each step is either on or off—with lesser values, the transitions are smoother.

The steps in the Trancegate were selected to bounce the sound across the stereo spectrum, and to give rhythm.

Finally, the sound runs into a 24 dB low-pass filter with the Cut-Off set to 812 Hz and the Resonance set to 44%. The cut-off frequency is then modulated by LFO1 where the Speed knob has been set to 0.08 Hz, and the Cut-Off knob has been set to 97. This setting allows the filter to gently shift over time.

T-Gate + Arp

T-Gate + Arp combines the Trancegate and Arpeggiator functions in Vanguard.

While the arpeggiator works best on chords, in this patch, you may want to try single notes. The reason for this suggestion is that, in my opinion, in this patch the arpeggiator functions better to restrict the length of notes than it does as an arpeggiator. However, please do try chords too.

As I said, the arpeggiator works on the length of the note, or its "gate" time. For this patch, I set the arpeggiator with the following settings:

- Mode: Up

- Speed: 1/16

- Gate: 1/32

- Octave: 1

I then set the Trancegate to 1/32 with the Contour set to 127, and set a pattern that (in conjunction with the arpeggiator) gives a strong rhythm with the sound bouncing across the stereo spectrum.

To finish the sound, I added some delay with the following settings:

- Mix: 41

- Type: Cross

- Time: 1/4

- Feedback: 21%

- Damp: 84

Playback Tools in Wusikstation

I have created three patches in Wusikstation. The basis for these three is 01 Vol Seq which takes advantage of the wave-sequence layer in Wusikstation, however, it does not actually sequence waves but uses the sequencer to control the volume. The following two patches, then add some pitch sequencing to allow Wusikstation to act like something between an arpeggiator and a sequencer.

01 Vol Seq

At its heart, this is a comparatively simple patch based on the Waldorf6 wave from the Famous Keys Volume 1 bank which is then filtered through a 2 pole low-pass filter with the Cut-Off frequency set to 36 and the Resonance set to 8. The sound is then pushed through a Quad Chorus and Stereo Echo to give the sound more dynamism. On its own that makes quite a dull sound. Two things give this patch some movement:

- First, the filter Cut-Off and the Resonance are modulated by two separate modulation LFOs which are running at different speeds so there is no "pattern" to the filter movement (although the LFOs are tempo synchronized).

- Second, the wave-sequence layer has been set to trigger the patch on every 1/16th note. A 16 step sequence has been set up, where each step lasts for 1/16 of a note. In the sequence lane, the Destination is the Volume of the layer, and each step has been set to a different volume, with the beats being of a higher volume. This is what gives the gated effect you hear.

02 Weird Pitch Seq

02 Weird Pitch Seq builds on 01 Vol Seq with a simple addition in the wave-sequencer.

A second sequence lane has been added with the Destination set to Tune. I then dragged my mouse over the steps to set a pattern. When I got something sufficiently odd, I saved it, and here we have the pattern. I suggest you only play single notes unless you want something *very* unmusical.

03 Arpeg Seq

While the last patch is interesting, it lacks something as a useful musical contribution. 03 Arpeg Seq rebalances some of that deficit.

This patch is also based on 01 Vol Seq, and has a second wave-sequence lane with the Destination set to Tune. However, rather than just set a few steps at random, this time I created an arpeggiated pattern.

In Wusikstation, you don't need to create patterns like I have chosen to do here—you can create a whole sequence where each note is different. You can set the steps in two ways:

- The more obvious way (but the harder way) is to drag each individual step to the desired level.

- The much easier way is by inputting the notes from a MIDI keyboard. To do this, right-click on the wave-sequence lane and select Seq Functions > MIDI Note Input. With each note you input, you will automatically then step forward to the next step.

Once I had created my arpeggiated pattern, I duplicated the whole layer to the W2 layer, raised the pitch of the original layer by an octave, and pushed the two layers slightly in opposite direction in the stereo spectrum.

As with 02 Weird Pitch Seq, you will find that this patch sounds best when you play single notes (or play octaves).

Playback Tools in Surge

The main playback tool in Surge is the Step Generator. However, this has a few useful tricks it can add:

- With the Deform control, the length (or gate time) of each Step can be controlled. This allows a more staccato sound to be created which will often emphasize the rhythmic properties of a sound.

- On a per-step basis, the filter and volume envelopes can be set to retrigger. This allows you to create more or less emphasis on the stepping process.

Both of these techniques are used in the sounds I'm about to create.

As with Wusikstation, you can use Surge's Step Generator to create pitched sequences. However, Surge doesn't have the Wusikstation's ready facility to input notes from a MIDI keyboard, so I haven't illustrated any of this type of patch here.

01 Pulsing 16s

As its name suggests, this is a patch which creates a pulsing 16th note rhythm. It is intended to be played as single notes in the lower region of the keyboard.

The basis of the sound is oscillator one and oscillator two. These both have the Classic oscillator loaded, I then tweaked the Shape, Width, Sub Width, and Sub Level controls to fatten up the sound, and then set the Uni Spread to 25 cents and the Uni Count to 7 voices. These oscillators are then panned left and right, and sent to the filter block which has been set up with the Wide configuration where only filter one is in operation. Filter one has its Cut-Off set at 300 Hz, and its Resonance set at 15%.

The rhythm is generated by the Step Generator which creates a 16-step pattern emphasizing beats 1, 5, 9, and 13 (in other words, the four beats in a 4/4 bar). This pattern is then set to modulate the cut-off frequency of filter one (the modulation is set just below 30). In the Step Generator LFO controls, the Rate is tempo sync'd at 1/16, and the Deform control is set to -40%.

As well as giving a changing tone through shifting the cut-off frequency, the effect of the Deform control is to help make the sequence more staccato and so give that pulsing effect. I could have turned the Deform control even lower but this tended to make the individual pulses much less distinct, so to emphasize the steps, I selected the envelope retrigger function for each step in the sequence. I then set the Amp EG as follows:

- Attack time: 0 seconds

- Decay time: 0.216 seconds

- Sustain level: 0%

To my ear, this setting gave sufficient emphasis to each step. If you want a more staccato sound, you could reduce the Decay time further.

I could have created a similar sound using a simple LFO, however, there are two key advantages that I have gained by using the Step Generator:

- First, I can set the level of each of the steps individually. In this case, this allows me to precisely sculpt the tone.

- Second, with Step Generator there is the option to retrigger the envelopes which gives me further sonic options.

02 Sequenced Noise

This next patch is one for Hawkwind fans. It's not an attempt to create anything musical, but rather a shifting noise patch. The patch doesn't use any oscillators, but instead, the noise source is used.

As a first step, the Noise Color control was set to its minimum, -100% which gives quite a "dark" noise color. This noise color was then modulated by an (LFO) envelope to transition the noise color from light to dark. This transition takes place over two bars (being controlled by the Decay time in LFO1 which is tempo-synchronized).

LFO one allows the tone slide in from bright to dark. LFO two, with the Step Generator engaged, then modulates the Noise Color on a continuing basis. The steps in LFO two were set on a fairly random basis, although every other step is louder. While the LFO modulates the Noise Color control to the maximum, the Magnitude slider was set to 36% so the color doesn't change too much.

As a final step, the LFO envelope was called in with the following settings:

- Delay time: 2 bars

- Attack time: 2 bars

- Sustain level: 100%

With this setting, the Step Generator starts to have its effect as the effect of LFO one's envelope is ending. Also, by setting an attack time, the effect of the Step Generator fades in.

To then round off the sound, two delay units were added. The first uses a chorus-type setting, and the second uses an echo set Left: 1/4, Right: 2/4.

03 Controlled Gating

You can, of course, allow the player to control the effect of the Step Generator. In 03 Controlled Gating, the effect of the Step Generator is controlled by the modulation wheel.

In essence, this patch is a classic synthesizer pad with a few tweaks. The foundation is two Classic oscillator sawtooth waveforms slightly tweaked and sent to the left and right channels. In the Filter Configuration, the Wide option has been selected to create a stereo sound.

In filter one, a 24 dB low-pass filter was selected with the Cut-Off set to 60 Hz and Resonance set to 28%. The Filter EG was then engaged with the following settings:

- Attack time: 2.4 seconds

- Decay time: 3.6 seconds

- Sustain level: 45%

- Release time: 6.9 seconds

- F1: 73.9 semitones

To complete the basic pad sound, the Amp EG was then set as follows:

- Attack time: 0.79 seconds

- Decay time: 2.1 seconds

- Sustain level: 63%

- Release time: 3 seconds

There are several areas of player control. The first comes with the second filter where a 24 dB high-pass filter has been selected with the Cut-Off and Resonance both set to the lowest amount (so the filter has no effect). These settings are then both controlled by the modulation wheel so that at its maximum setting the Cut-Off and Resonance are both raised giving a noticeably thinner and brighter sound.

The second area of player control comes with the Step Generator in LFO one. Before I talk about this player control, let me explain how this Step Generator has been set up.

With the examples so far, the Step Generator has been used to modulate "up". In other words, the modulation destination has been set to its minimum and then the modulation source works to increase that minimum. In this case, the Step Generator modulates the Output Volume, but modulates down. This modulation is quite simple: the Volume is set, and then the modulation assign mode is engaged and the Volume control is moved down, rather than up.

In the Step Generator, there are two groups of four steps. The four steps fill every other slot, and are of decreasing magnitude. The Rate has been tempo sync'd to 1/16 and the Deform control has been set to 61% so each step lasts slightly longer than its assigned time length. Because the values of the steps fall, this means that the amount by which the Volume is reduced falls, so the volume of the volume-reduced steps increases. In the steps where no value has been inserted, the volume is heard at its full amount.

The last tweak I made to the Step Generator LFO was to set the Magnitude slider to 0%. This means that the Step Generator has no effect on the Volume. Now, before you start wondering about an apparently pointless Step Generator, you will remember that my intention was to give the player some control, so the next thing I did was to assign the modulation wheel to also control the Magnitude slider.

This assignment means that when the modulation wheel is pushed, the Magnitude is increased and only then does the Step Generator begin to affect the Output Volume. You will also remember that the high-pass filter's cut-off and resonance is also affected by modulation wheel. The combination of the two gives the player considerable control over this effect.

As a final step, I added an echo with the Left delay set to 1/16 and the Right delay set to 1/4. This echo works when the modulation wheel is at rest, and when it is pushed. When the wheel is pushed, and so the filtered gating effect is heard, the echo works to emphasize this sound, giving a gated effect bouncing around the stereo spectrum.

Pushing the Sonic Envelope

The sounds that I have created so far in this book have been designed with a specific purpose, and sculpted to fit within a musical context. Designing sounds that fit within these parameters means that the sounds may not have many applications outside of those musical contexts. For this refill, I want to break away from some of those restrictions and create some sounds with a broader sonic texture.

Weird Stuff in Z3TA+

As we've already seen throughout this book, Z3TA+ is an immensely powerful synthesizer. As you would expect, as well as those controlled sounds that we have already created, it is capable of going wild. I'm not going to push it anywhere near to its limits, but in this refill I do want to take a step away from some of the musically useful sounds that we have created thus far.

Dominant Attack

The first patch I am going to create is the most useful musically of this group. The challenge with this patch, and hence its name, is that it will dominate the sonic spectrum, perhaps to an unhealthy extent. As you will hear, this is a very up front sound which could be useful in many contexts, however, you are likely to have to use some serious amounts of EQ, or to put it into a very sparse track, if you want it to sit well.

In creating this sound, I was trying to create a keyboard-esque sound. You will hear that this sound has elements of an electric piano, and perhaps a harpsichord. It also has a synthetic quality of its own.

The basis for this sound is created with oscillator one where a Vintage Saw 1 wave has been loaded. I then tweaked this with the Waveshaper, pushing the Warp and Twist controls a bit to give a thinner, sharper tone which has a buzzy quality to it. I immediately addressed the buzzy quality by pushing the wave through a 12 dB low-pass filter. For reasons that will become clearer, I balanced this oscillator between the two filters and then set the Cut-Off in each filter to 16 Hz, and the Resonance to 2.3 dB.

Clearly this amount of heavy filtering flattened the sound and so I set Envelope one to control the filter with the following settings in the modulation matrix:

- Source: EG1

- Range: Minimum 50%, Maximum 100%

- Control: Velocity

- Destination: All Filters Cut-Off

Envelope one is then set as follows:

- Attack Time: 0

- Slope Time: 0.15 seconds

- Slope curve: exponential (convex)

- Slope Level: 75%

- Decay Time: 2.9 seconds

- Sustain Level: 15%

- Release time: 0.02 seconds

- Amount: 100%

The combination of the envelope and the Velocity control gives a very playable, almost clavinet-like tone, but I'm after a much more immediate, brighter sound. The first step to brightening the tone was to copy oscillator one to oscillators two and three—as the Waveshaper has been used, the wave shape needs to be separately copied. Once copied, oscillators two and three were raised by an octave, and panned left and right respectively. The levels of the new oscillators were then also dropped to 60%.

These two oscillators give a brighter sound, but it still didn't have enough attack for what I was after. To get this tone, I loaded oscillators four and five with a sine wave and set oscillator four to frequency modulate oscillator five, and pushed oscillator five's Level to 100%. With the modulation set up, I set oscillator four's Octave to +3, and Transp to +7, and oscillator five's Octave to +1, so the modulator is two octaves and seven semitones above the carrier. In the modulation matrix I then set Envelope One to control oscillator four's Level, with Velocity as the Control.

This FM addition gave me the amount of bite I was looking for. However, it also exposed some of the lack of weight in the lower regions of the key range. To rectify this, I called up a Vintage Square 1 wave into oscillator six. I pitched this down an octave and set the Level (by ear) to 60%. This gave me the tone I was after, but was rather too dominant in the higher keyboard ranges so I set some key scaling in the modulation matrix:

- Source: On

- Range: Minimum 0%, Maximum 80%

- Curve: U-Lin-

- Destination: Osc6 Level

Finally, for a patch with this much attack, I switched on the Limiter.

Dominant Screamer

Dominant Screamer has been designed as a rather off-the-wall lead-type sound. By no means will it work in all situations where you are looking for a lead sound!

Like Dominant Attack, the foundation of Dominant Screamer is a Vintage Saw 1 wave which has been mangled in the Waveshaper. In this instance, I tweaked the Offset, Selfsync, and Bit Reduction controls to get the sharper, brighter waveform that I was after. Having created the wave, I then copied oscillator one to oscillator two without making any changes, so there's no detuning etc, only an increase in volume that comes with a doubling of oscillators.

As a final tweak in the oscillators' section, I turned to the Global page and set Bend Up to +5 and Bend Down to -12 so I can push this note's pitch over a wide frequency range.

The oscillators are both bussed to be split between the two filters (which are arranged in parallel). In the filter block, I set both filters with a 12 dB low-pass filter and put the Cut-Off control at 129 Hz, and the Resonance control to 2.9 dB. I then panned filter one to the left and filter two to the right. At the moment this panning has no effect since the oscillators are both balanced between the filters.

At this point, the filter settings have taken away most of the tone from the patch, so I introduced some velocity control using the following settings in the modulation matrix:

- Source: On

- Range: Minimum 0%, Maximum 83%

- Control: Velocity

- Destination: All Filters Cutoff

This gave me a broad range of tone, but didn't give me the scream I was after. The scream I ended up with came through detuning oscillator two. Instead of having a permanent detuning, I wanted to put the scream control into the hands of the player, so I hooked up the pitch bend to also control the detuning. This means that as a note is bent with the pitch bend wheel, the second oscillator is also detuned relative to oscillator one giving the scream. I used the following settings in the modulation matrix to achieve this effect:

- Source: On

- Range: Minimum 0%, Maximum 65%

235

- Curve: Pitch 1S

- Control: U-Bend

- Destination: Osc2 Pitch

At higher velocities, and in higher keyboard regions, this effect gives a sound somewhere between a scream and a siren. However, I wanted to give slightly more audio emphasis to the scream effect.

The first change I made to emphasize the effect was to give the sound some physical movement in the stereo spectrum. As the scream increases, in other words as the pitch wheel is pushed, oscillator one is panned to the left and oscillator two is panned to the right. Also, as the scream increases, an echo is added.

To pan oscillator one, I set the following line in the modulation matrix:

- Source: On

- Range: Minimum 0%, Maximum 100%

- Curve: B-Lin+

- Control: B-Bend

- Destination: Osc1 Bus

The settings to pan oscillator two are the same, except the Curve is B-Lin-, and the Destination is Osc2 Bus.

For the echo, I used the following settings:

- Mode: LRC Delay

- Sync: 1/2

- EQ Mode: Wide

- Feedback: 82%

- Low: -20 dB

- Mid: -6 dB

- High: -10 dB

- Level: 85%

Once the echo was in place, I used the following modulation matrix settings to link the echo to the pitch bend:

- Source: On

- Range: Minimum 0%, Maximum 100%

- Control: U-Bend

■ Destination: Delay1 Level

This setting means that the echo is only heard when the pitch bend wheel is pushed.

Dirty Sustain Bass

I wanted to create a growly, distorted, foghorn sort of bass sound to use as a drone. There are many ways to create this sort of sound. With Dirty Sustain Bass, I took one option—you will be able to find many other ways of creating a similar sound. The advantage of the method I have chosen here is that you get a subtle constantly shifting tone to the sound.

This sound uses ring modulation in a controlled manner. With the next patch, Dissonant XY Pad, we will hear ring modulation used to create slightly wilder tones. Dirty Sustain Bass is created with two pairs of ring modulating oscillators. All four oscillators have a Vintage Saw 1 wave loaded. The two carrier oscillators, oscillator two and oscillator four, then have their outputs sent to filter one.

In the first pair, oscillator one has its Octave set to +1 and its Transp set to +7. Oscillator two does not have its pitch changed. The second pair is pitched an octave below the first pair, so oscillator three has its Octave set to 0 and its Transp set to +7, and oscillator 4 has its Octave set to -1.

While these ring modulated oscillators give a fairly bright, aggressive sound, they don't give a sufficiently dirty sound. To remedy this, I put a 24 dB low-pass filter into filter one and set the Cut-Off to 16 Hz, and the Resonance to 2.3 dB. This flattened the sound so I added two lines in the modulation matrix. The first line opens up the filter as higher notes are played:

■ Source: U-Note #

■ Range: Minimum 0%, Maximum 60%

■ Destination: Filter1 Cutoff

The second line adds some velocity sensitivity to the filter cut-off:

■ Source: On

■ Range: Minimum 2.7%, Maximum 34%

■ Control: Velocity

■ Destination: Filter1 Cutoff

The combination of these two lines gave me sufficient control over the tone of the patch. However, I still wanted to give the player some control over the movement of the underlying sound, so I called up LFO one and set it as follows:

■ Wave: Sine

■ Phase: Free

■ Speed: 0.25 Hz

I then hooked up the LFO so that it would gently modulate the pitch of the two modulating oscillators (oscillator one and oscillator three). The effect of changing the pitch of the modulators in relation to the carrier is to shift the tone rather than to change the pitch. I used the following settings in the modulation matrix to achieve the modulation:

- Source: LFO1

- Range: Minimum 0%, Maximum 15%

- Curve: Pitch 1S

- Control: Mod Wheel

- Destination: Osc1 Pitch (and Osc3 Pitch for the second modulator)

I did try this patch with a separate LFO modulating oscillator one and oscillator two. This lost some cohesion in the sound, so I kept to a single LFO.

Dissonant XY Pad

As you've probably guessed from the name of this patch, Dissonant XY Pad uses the XY pad in Z3TA+, so as a first step, open up the XY Pad. Here's something you might not know: hold a note and then right-click on the XY pad and you will see the cross-hairs jumping about randomly. This should give you some idea of the variations of tone that are available in the patch. If you want a somewhat less random sound, then left-click on the XY pad and drag the cross-hairs around.

Let me explain how this sound was created.

With this patch, I was looking to create some metallic ring-modulated sounds, and to then combine these results with a resonant filter to mangle the sounds still further. I did not intend to create anything musically useful (as the results will testify).

To create the ring modulation, I loaded a Vintage Saw 1 wave into oscillator one, set the Transp control to +1, and selected the Ring Group. In oscillator two, I loaded a Sine wave and pushed the bus slider so that the oscillator's output goes to filter one. The Transp control in oscillator one isn't particularly significant—I chose this value as it would ensure there was a dissonant sound from the outset.

In the filter, I selected a 12 dB low-pass filter and set its Cut-Off to 250 Hz and Resonance to 14 dB.

To then create the wackiness, I opened up the XY pad and hooked it up through the modulation matrix. Firstly, I set the horizontal axis (the X axis) to control the pitch of oscillator one—this affects the ring modulation giving a wide range of tones. I used the following settings:

- Source: On

- Range: Minimum 0%, Maximum 100%

- Curve: Pitch 4O

- Control: X-Y Pad Y

- Destination: Osc1 Pitch

Using the Pitch 4O curve, I ensure that the X axis allows for a wide range of tonal shift.

I set the vertical axis (the Y axis) to control the filter cut-off and so shape the tone further with the following settings:

- Source: On

- Range: Minimum 0%, Maximum 50%

- Control: X-Y Pad Y

- Destination: Filter1 Cutoff

Now go and see what tones you can find...

Weird Stuff with Surge

Well... not necessarily weird stuff with Surge, but here are a few different ideas for you to think about. First off, I want to take another look at the notion of FM wave-sequencing. This idea was first discussed in Chapter 8: Wave-Sequencing.

01 Chaos Sweep

The basis of 01 Chaos Sweep is a simple FM stack with a Sine wave in oscillator 1 (which acts as the carrier) and the Waldorf > Chaosweep wavetable wave in oscillator two (which acts as the modulator. To create the FM sound, the FM Depth is set to -16 dB.

If you've read the rest of the book—and if you haven't that probably means you are reading the free refill, and so I urge you to go and buy the book: there's a lot of good stuff in it!!—you will know that in Surge the wavetable oscillators are not single waveforms. Instead, they are a series of related selectable (and manipulable) waveshapes. The basic Chaosweep wave shape didn't quite give me the sound I was after, so I pushed the Shape control in oscillator two to around 50%. At this point I got the sound I was looking for.

However, I found this shape a bit to static for my taste, so I decided to mess with things a bit, and called up LFO one which I set to modulate oscillator two's Shape. LFO one's modulation will cause oscillator two to sweep through its available waveforms. Since oscillator two is the frequency modulator of oscillator one, this will cause a constant shift in tone.

To make the initial effect of the wave-sweep (which in turn affects the frequency modulation) very subtle, I arranged the LFO with the following settings:

- Wave: Sine

- Rate: 0.021 Hz

- Magnitude: 14.5%

By setting the Magnitude to this low level, I ensured that the wave-sweep only passes through a limited number of waves. If you hold and sustain a note, you will just about hear the wave-sweep. At this point I decided to give the player more control and so I called up the modulation wheel modulation settings and assigned it to:

- the LFO Rate, and

- oscillator two's Saturate control.

The combination of these two modulations means that as the modulation wheel gets pushed, the LFO rate increases which subjectively increases the prominence of the wave-sweep. In addition, the Saturate control is increased giving a more distorted tone to the modulator, and hence more character to the frequency modulated sine wave in oscillator one.

You will notice that all of this tone shifting is accomplished without resorting to the filters.

02 Manual Bass Echo

On first hearing, 02 Manual Bass Echo seems like a quite straightforward, if somewhat rich, synth bass sound. However, this sound is actually two layers—one in Scene A, and the other in Scene B. In Scene A, you have the synth bass sound, and in Scene B, there is a manually created echo. The prominence of that echo can be controlled by the player.

The echo is unlike a normal echo in that it is an echo on the initial impact of the note—if you sustain a note, then the echo will dissipate before the note is released. However, if you call up the echo and play a staccato part, then the echo will become prominent.

I'll look at the two parts separately, and then look at how the two elements can be pulled together.

In Scene A, there is the main synth bass sound. I quite like this as a synth bass sound—it is quite warm, rich, and prominent. It is created around two Classic oscillators with the sawtooth wave selected. I then messed with the Width, Sub Width, and Sub Level controls, and set Uni Spread to 10 cents and Uni Count to 7. Oscillator two is a copy of oscillator one, but is slightly detuned.

Oscillator one is then sent to the left and oscillator two to the right. The Wide filter configuration was then selected to keep this stereo panning. To complete this first triumvirate of oscillators, in oscillator three I loaded a Classic oscillator and selected the square waveform, and then dropped the pitch by an octave. I didn't use the unison/multimode options and kept the panning to the center.

Only one filter is used in this scene: filter one where the low-pass Ladder filter was selected with mode 3 engaged. I set the Cut-Off to 56 Hz, the Resonance to 22%, and then set the Filter EG as follows:

- Attack time: 0 seconds

- Decay time: 0.446 seconds

- Sustain level: 53%

- Release time: 0.25 seconds

- F1 slider: 59 semitones

To complete this scene, I set the Amp EG as follows:

- Attack time: 0 seconds

- Decay time: 2.3 seconds

- Sustain level: 46%

- Release time: 0.26 seconds

Finally, to thicken up the sound a bit, I engaged the Soft waveshaper and set the slider to 5.4 dB.

Once Scene A was set, I copied it to Scene B and started tweaking.

The first two tweaks I made were to:

- set the HP (high-pass) slider to 300 Hz to thin out the sound considerably, and

- push the waveshaper slider up to 13 dB to make the sound considerably dirtier.

I then made a few tweaks to the filter by engaging mode 4, and setting Cut-Off to 13 Hz. The Resonance setting was left unchanged. The filter envelope was then set:

- Attack time: 0 seconds

- Decay time: 3 seconds

- Sustain level: 0%

- Release time: 15 seconds

- F1 slider: 96 semitones

The Amp EG was also tweaked:

- Attack time: 0 seconds

- Decay time: 0 seconds

- Sustain level: 100%

- Release time: 16 seconds

I then called up filter 2. This filter works like a volume control. I selected the same low-pass Ladder filter in mode 4 that is in filter one, and set the Cut-Off and Resonance to their lowest settings. To control filter one's cut-off I called up LFO one and selected the envelope option with the following settings:

- Delay time: 0 seconds

- Attack time: 0seconds

- Hold time: 0 seconds

- Decay time: 2.63 seconds

- Sustain level: 0%

- Release time: 2.63 seconds

I then set the LFO to fully modulate the cut-off frequency. With this setting, when the note is struck, it decays quite rapidly. I then wanted to create the echo type effect for which I called up LFO 2 with a Step Generator in which I put a step in every other space—each step progressively decays from the previous. Then I tempo sync'd the Rate to 1/8, adjusted the Scene Volume to 0, and set this LFO to modulate the volume. When a note is struck, this set up gives a decaying echo effect.

To then link the sounds together, I engaged Dual mode. To control the volume of Scene B by the modulation wheel, in Scene B, I:

- went to LFO1 and set the Magnitude slider to 0

- engaged modulation assign mode for the Modwheel, and

- set the modulation wheel to fully control the Magnitude, so if you push the modulation wheel to its maximum value, the Magnitude control will operate at its maximum value and the effect of the echo can be heard.

03 Big Bass

This final patch in this refill shares a similar foundation to Scene A from the last patch, 02 Manual Bass Echo. This patch develops the sound further, but takes it to a certain extreme: you may find it is quite hard to get a sound like this to fit in a mix.

The oscillators are the same as in the last patch, and the filter configurations are the same. The waveshaper is also the same. However, the filters and the envelopes are tweaked to give quite a different tone.

The first change from the last patch that you may notice is that this sound has much less attack. It still has a very swift attack, but this patch has much more of a low-end thump, rather than a mid-range twang. This change in emphasis is due to a change in the filter envelope.

In filter one, I set the Cut-Off to 59 Hz and the Resonance to 21%. I then set the Filter EG as follows:

- Attack time: 0.017 seconds

- Decay time: 0.446 seconds

- Sustain level: 53%

- Release time: 0.25 seconds

- F1 slider: 38 semitones

The volume envelope is set as for the previous patch.

The result of the tweaks is a much thicker sound, with an almost muted-brassy texture.

Scene B is very similar to Scene A. One key difference is that the levels of oscillators one and two have been dropped to around -5 dB. The other key changes are to the filter, and the two envelopes.

In filter one, the Cut-Off has been set to 27 Hz, and the Resonance has been set to 27%. The filter envelope was then set as follows:

- Attack time: 0.017 seconds

- Decay time: 0.425 seconds

- Sustain level: 53%

- Release time: 1.5 seconds

- F1 slider: 47 semitones

The volume (Amp EG) was then set:

- Attack time: 0.16 seconds

- Decay time: 2.2 seconds

- Sustain level: 45%

- Release time: 1.8 seconds

The effect of these changes is that Scene B is less predominant at the start of a note, but can be clearly heard during the release stage.

To then finish this sound, I set the Volume of Scene B to zero, and set the Velocity to modulate the volume (so maximum velocity increases the volume to its maximum level). This ensures that there is a thick sound but louder notes have more "ring" to them due to the emphasis on the slower release section.

Chapter 12—Refill #3
Synthetic Drums and Percussion

This third refill focuses on synthetic drums and percussion. I am going to build all of these sounds with Z3TA+.

There are a couple of things I want to get straight at the start:

- First, just because this section covers drums and percussion, don't skim over this section. There are several techniques that are applicable to more conventional sounds.

- Second, the key word in this refill's title is "synthetic". The sounds I am going to create here are intended to sound like electronic drums (but are not necessarily intended to sound like a piece of analog hardware). If you want realistic acoustic sounding drums, then either hire a real drummer or go and buy yourself a set of high quality samples.

However, if you are going to use samples, then you might want to come back and check out some of the techniques here—you will often find that layering samples and synthetic drums can keep the acoustic sound but give the necessary depth and power which is often missed when using samples alone.

Background to Creating Synthetic Drum and Percussion Sounds

There are specialized drum tools—for instance DR-008 from FXPansion (www.fxpansion.com), RMIV from Linplug (www.linplug.com), and Microtonic from Sonic Charge (www.soniccharge.com)—which can help you create synthetic drum sounds. However, unless you understand what you're doing, then it is hard to use these specialized tools. Once you've been through this refill and have got a grip on synthetic percussion, you might want to take a look at these specialist tools.

You may also find that these specialized drum tools have a limited functionality: that is the nature of specialized tools. If they weren't dedicated to a specific purpose, then they would be general purpose synthesizers. All of the synthesizers featured in this book are more flexible than these specialized drum tools, however, they do have limitations when compared with the drum tools.

The main limitation is that (apart from Wusikstation), none of the synthesizers is multi-timbral. The drum tools are set up so that you can have a number of different sounds (for instance, a kick drum, some hats, a clap, and a snare) all playing together. With the synthesizers in this book, you will need a new instance of a synthesizer for each different sound. The flip-side is that your individual drum and percussion sounds will be playable over a much wider keyboard range, giving far more tonal variation.

Most synthetic drum sounds have two key elements:

- a sound, which is often a mangled sine wave, and

- some noise, which is usually heavily filtered.

Rather than trying to explain the theory, it's probably easiest if we leap straight in and start making some drum sounds. The intention behind all of the sounds in this refill is that they are quite raw—once you have got your head around the concepts I am demonstrating, I suggest you take some time to finesse these sounds and perfect them for use in your own productions.

For all of the sounds in this refill, I have set the Poly control to 4 so that any rolls do not cause an early cut out of the sustained portion of a drum sound. I have also set the Render mode to High in order to get a clearer sound without any unwanted audio artefacts.

Kick Drum

The essence of a kick drum is a low-pitched sine wave. This sine wave is then controlled by two envelopes:

- A volume envelope gives the characteristic staccato drum sound.

- A pitch envelope gives the sound of the "beater" "hitting" the "drum".

If you open up the Kick Drum patch and have a look, you will see that a Sine wave has been loaded in oscillator one. This wave has had Fixed, Free mode selected so the pitch of the wave is not affected by the MIDI input pitch. The note has then been dropped by two octaves to sound at a suitable pitch. The sine wave is sent to the right bus (that is to filter two) where no filter has been selected so the clean sound passes to the output.

Before the sine wave reaches the output, it passes through the Amplifier envelope which has been set to give a classic drum-type volume envelope, with the following settings:

- Delay Time: 0 seconds

- Attack Time: 0 seconds

- Slope Time: 0.02 seconds—in conjunction with the Slope Level, this introduces a hold stage to give the drum slightly more kick

- Slope Level: 99%

- Decay Time: 0.58 seconds (with an exponential curve selected)

- Sustain Level: 0%

This gives a reasonable sort of drum sound, but it lacks real kick. To put the punch into this drum, I added a sharp bit of pitch modulation to the sine wave. This is a great technique for adding impact to the start of notes—you can use it with many types of sounds, not just drums. For instance, it works well with bass sounds.

(Unsurprisingly) I used the Pitch Envelope to modulate the pitch of the sine wave. There are two relevant settings in the Pitch Envelope:

- Start Level is set to 100%, and

- Attack Time is set to 0.03 seconds

The modulation is then hooked up through the modulation matrix where the following settings where dialed up:

- Source: Pitch EG

- Range: 0% to 58%

- Curve: Pitch 4O

- Control: Velocity

- Destination: Osc1 Pitch.

You will notice that Velocity has been set as the Control. This allows the effect of the Pitch Envelope to be controlled by velocity, thereby giving a more flexible tone but maintaining a reasonable volume level over the whole velocity range.

I set this drum to give a good all-round sound. However, there are several things you can do to tailor its sound to your particular purpose.

First off, the drum sound is quite "hard". If you wanted to soften it a bit, there are a few things to think about. For instance:

- You could slow down the Attack Time in the Amplitude envelope. Clearly, if you slow it too much then the sound will fade in, which is not what you want for a drum. However, if you push the slider to around 0.01 seconds, then you will hear a noticeable softening to the attack without hearing a fade in.

- In conjunction with this, you could also increase the Decay Time in the Amplitude envelope. This will mean that the sound decays more slowly and so you will hear a greater proportion of the tone when the pitch envelope has ceased to have effect. In a mix, this is unlikely to be noticeable. However, what you will notice is a further "softening" of the sound.

247

■ Another significant change you can make is to the Range control. If you reduce the amount by which the Pitch Envelope modulates the pitch of oscillator one, then you will soften the tone, reducing that initial sound of the "beater" hitting the drum. If you also adjust the Attack Time in the Pitch envelope, you will hear that it varies the tone of the beater. This tone change can work well in conjunction with the different volume envelope.

Conversely, you might want to get a harder drum sound. There are several things you can do here but the most immediate change is to increase the Range setting so that the Pitch Envelope modulates the pitch of oscillator one to a greater extent. In conjunction with this you can adjust the Attack Time in the Pitch envelope, although you may find that without cautious adjustment you get a syn drum-type sound (which I will demonstrate later in this refill).

Once you have pushed up the Range control, the other main change you might want to consider is to increase the Slope Time in the Amplitude envelope. This will give more prominence to the attack portion of the drum hit.

To slightly round out this drum sound, I then added a second oscillator. This is quite subtle— adding weight more than tone, and you may not notice or like the change, however, it is a choice that I liked, so I will tell you what I did.

First I loaded a Vintage Pulse 1 wave into oscillator two. I then dropped its pitch by two octaves. Unlike oscillator one, the pitch of this oscillator responds fully to the MIDI input data. I then sent this oscillator's output to the left bus (filter one). You remember that the sine wave in oscillator one is sent to the right bus.

In the filter section, I selected a 24 dB/octave low-pass filter and set the Cut-Off slider to 83 Hz and the Resonance slider to 17 dB. This gives a very muted sound. The level of this sound is then modulated by Envelope One, with Velocity set in the Control column.

If you want to make this sound more prominent—so making the sound something of a cross between a kick drum and a low bass—then you could add the Pitch 40 option in the Curve column and adjust the Range to taste.

Closed Hats

Synthetic hi hats are quite straightforward: they are essentially a bit of filtered noise. I'll start by looking at a closed hi hat.

For this patch I loaded the Noise (Octave) waveform into oscillator one and changed the Octave setting to +5. The noise source used here is quite dirty. By increasing its pitch by five octaves, I got a much cleaner form of noise that (to my ear) sounds closer to white noise. This is more appropriate for the sound I am after (and it's also a lot easier than saying play this patch at the very top of your keyboard).

To further shape the tone of a closed hi hat, I directed the noise source to filter one (by pushing the Bus slider to the left bus). In filter one I selected the 12 dB/octave band-pass filter. I felt that setting the Cut-Off frequency to around 1,461 Hz gave me the ideal tone.

However, this still sounds like a lot of noise and not a hi hat. The final piece of the jigsaw is the envelope to control the volume. Here I used the Amplifier envelope where only two setting matter: the Slope Level which is set to 100% and the Decay Time which is set to 0.11 seconds (with an exponential, that is concave, slope). Everything else is set to zero (apart from the Amount control which is also set to the maximum).

With these level settings, I got the hi hat character that I was after. To then give a bit of volume sensitivity to the patch, I hooked up the modulation matrix with the following settings:

- Source: On

- Range: Minimum 46%, Maximum 100%

- Control: Velocity

- Destination: Oscillator 1 Level

With the raised minimum setting, this is quite a subtle modulation, and it doesn't allow for any change in tone. If you do want to change the tone, there are two things you might consider:

- The cut-off frequency in filter one could be modulated (very gently) by an LFO. This could give a subtle variation in tone each time a note is triggered.

- If you want more control over the tonal variation, you could again modulate the cut-off frequency, but use pitch as the modulation source. In this way, you could program pitch variation in your MIDI track.

You could, of course, use both of these techniques.

Open Hats

Having set up the closed hi hat, the next logical step is the open hi hat. This patch was built on the closed hat, but with two tweaks.

- The first change was to select the 24 dB band-pass filter. This filter gives a slightly thinner tone which means the patch is less dominant (which is a good thing, in light of the next tweak).

- The second change was in the Amp envelope where the Decay Time was set to 1.32 seconds to mimic the longer decay of an open hi hat.

Clearly, I could have made many other changes—for instance, I could have dialed a different cut-off frequency—however, if I made too many radical changes the open hat may not have worked with the closed hat. Although this is a synthetic percussion sound, I think it is important that open and closed hats sound as if they may be related.

Snare Drum

If you've got the patches that accompany this book, you'll see that I've called this Dodgy Snare. This is not a conventional synthesizer snare—instead it is quite low pitched and fat.

There are two elements in this drum sound:

- The drum sound itself. This is created with two sine waves in oscillators one and two.

- The snare sound which is created by the Noise (Octave) wave.

Let's look at these three oscillators in turn.

In oscillator one, which creates the main drum sound, I have loaded a Sine wave. The Mode of this oscillator has been set to Fixed, Free meaning that the oscillator does not respond to MIDI pitch information. Instead, this oscillator has been tuned by setting Octave to -1 and Transp to -6. This oscillator is bussed to filter 2 where no filter has been selected, and so the raw oscillator is directly output.

To then give this sine wave some shape, and give it some of the characteristics of a snare drum, I set EG1 to modulate the Level of oscillator one. Everything in Envelope One was set to zero apart from the following:

- Slope Level: 100%

- Decay Time: 0.15 seconds

- Release Time: 0.02 seconds

This gives us a sine wave with a snare-like volume envelope. To give the sound a bit more impact, I called up the Pitch Envelope and used it in a similar way to the way it was used in the Kick Drum patch earlier in this refill.

In this case, in the Pitch Envelope:

- Start Level was set to 100%

- Attack Time was set to 0.01 seconds

- Amount was set to 100%

The Pitch Envelope was then hooked up through the modulation matrix with the following settings:

- Source: Pitch EG

- Range: Minimum 0%, Maximum 83%

- Curve: Pitch 40

- Destination: Osc1 Pitch.

As you can hear, you can adjust the tone of this modulation by fiddling with the Range setting.

I then wanted to thicken up this sound a bit and add some lower-pitched ring. As a first step in doing this, I:

- copied oscillator one to oscillator two

- copied EG1 to EG2, and set EG2 to modulate the level of oscillator two

- set the Pitch EG to modulate Oscillator two's pitch.

I then made a few tweaks:

- in oscillator two, Octave was set to -2 and Transp to +3, pitching this sine wave below oscillator one

- in Envelope two, the Decay Time was set to 0.29 seconds, giving more ring to the sound, and

- in the modulation matrix, the Range Maximum was set to 100%.

The cumulative effect of these three tweaks is to create a much lower ringing tone which slightly interferes with oscillator one. As a final step, I set the Level in oscillator two to 22% which was the point where the second oscillator added tone but could not be separately identified.

The final part to this snare drum is the noise which provides the initial drum "crack". As I did with the hi hats, I used the Noise (Octave) and pushed the Octave setting to +5. The Level of oscillator three is controlled by Envelope three where the Slope Level is set to 100% and the Decay Time is set to 0.15 seconds. I also pulled down the Level control to 32% so that it worked in the patch—again, my aim here was to set the oscillator at a level where I got a cohesive sound.

The final bit of tone shaping that I applied to this noise source was to send it to the filter (the Bus setting in Oscillator three is set to the left, in other words, filter one). I then set filter one as follows:

- Type: 12 dB band-pass

- Cut-Off: 2,000 Hz

- Resonance: 35 dB

You can adjust the tone of the "crack" to taste with the cut-off and resonance controls.

A couple of things you might notice. First, I didn't do anything with the Amp (volume) envelope. If you've got the patches, you will see it is set, however, it doesn't have any affect on this sound. Also, you will note that there is no real-time tone or volume control in this patch. If this upsets you, feel free to get creative.

Rim Shot

The Rim Shot is a very swift patch based on a single oscillator with the Noise (Octave) waveform.

For this patch I set the Octave control to +2 and then ran the oscillator into a 12 dB low-pass filter with the Cut-Off set to 2,000 Hz and the Resonance set to 28 dB. This gave me a tone that I like.

I then set the Amp envelope to give the volume the shape of a rim shot with the following settings:

- Slope Level: 100%

- Decay Time: 0.18 seconds

- Decay curve: exponential (concave)

- Amount: 100%

You can further adjust the tone of this sound by playing with the octave control as well as the cut-off and resonance controls.

Claps

As you might have guessed by now, the sound of clapping is created with the Noise (Octave) wave. Here I have loaded that wave into oscillators one and two. Oscillator one has its Octave set to +5 and is then directed to filter one, and oscillator two has its Octave set to +4 and is then directed to filter two.

In the filter section, 12 dB band-pass filters are selected. Filter one's Cut-Off is set to 2,034 Hz, and filter two's Cut-Off is set to 1,822 Hz. After this, the Level of oscillator one was increased to 100% and the Level of oscillator two was pushed to 80%.

While these two oscillators create something of the tone of hand claps, they ignore a key characteristic. Often the sound you are after will not be a single hand clap, but the sound of many people clapping together. Unlike, say, a string section, when many people clap at the same time you don't get a chorusing effect. Instead you get something between a tremolo and an echo effect: lots of individual impacts very close together.

We could replicate this with a short echo. However, this is likely to create chorusing or comb-filtering type effects, neither of which is desirable. Instead, I created the effect with a low frequency oscillator.

I called up LFO one, loaded a Pulse 50% wave, and pushed the Speed up to 20 Hz. In the modulation matrix, I set this LFO as the Source with oscillator one's Level as the Destination, and the Range set to full. I set the LFO to only affect oscillator one so this effect doesn't get too extreme.

However, at this LFO speed, the sound I was getting was too much like a tinny echo. To rectify this, I increased the speed of the LFO. You can't simply increase the speed by pushing the Speed slider since it is already at the maximum. Instead, I added another line in the modulation matrix with the following settings:

- Source: On

- Range: Minimum 0%, Maximum 87%

- Curve: Pitch 40

- Destination: LFO1 Speed

For this set up, it is the Curve setting that allows us to increase the LFO speed. As you can hear, this is a very subtle effect. It is more noticeable if you mute the second oscillator. However, as you increase the Range control, you can hear the echoes become less pronounced and the sound of a number of a number of people clapping is heard instead.

Cowbell

Now, take a deep breath... You're in for a shock. For this cowbell sound I'm going to use a different waveform.

For this cowbell sound, I have loaded a Vintage Square 1 wave into oscillators one a two. The Octave is set to +2 in oscillator one, and in oscillator two the Octave is set to +1 and the Trasp to +5. Both oscillators are then fed into filter one where a 12 dB low-pass filter has been selected with the Cut-Off set to 605 Hz, and the Resonance set to 28 dB.

To complete this sound, I set the Amp envelope with the following settings:

- Slope Level: 100%

- Decay Time: 0.34 seconds

- Decay curve: exponential

- Amount: 100%

I completed this patch by switching on the Limiter.

You will notice that this sound is fully pitch scaled. I suggest you play it around the middle of your keyboard for best results.

Syn Drum

If you don't know what a syn drum sounds like, then you weren't around in the 1970s. In this case you probably also missed out on disco, which may be no bad thing.

Like the cowbell sound we just created, this patch is playable across the whole keyboard. In my opinion, this sound is best in the lower-middle of the keyboard.

The basis of this sound is a sine wave in oscillator one with the Octave set to -1. This is then modulated by the Pitch EG which has the following settings:

- Start Level: 100%

- Attack Time: 0.77 seconds

- Amount: 100%

This is then hooked up to oscillator one through the modulation matrix with the following settings:

- Source: Pitch EG

- Range: Minimum 0%, Maximum 100%

- Curve: Pitch 40

- Destination: Osc1 Pitch

You will notice that the Attack Time in the envelope is quite slow (at 0.77 seconds). The intention behind this setting is to ensure that the falling pitch is heard. If this time is too short, then the Pitch envelope will work to give an impact as it does in the Kick Drum patch.

The last step in creating this patch is the volume envelope. As you might expect, I used the Amp envelope with the following settings:

- Slope Level: 100%

- Decay Time: 1.47 seconds

- Decay curve: exponential

- Amount: 100%

Toms

The last patch I am going to create is a tom sound. As with the syn drum, this patch is playable across the whole keyboard. You will probably find the most convincing synthetic tom sound around the middle of your keyboard.

The main body of this sound is built around a sine wave loaded into oscillator two. The sine wave is fed into filter one, although this filter has little effect on the wave (the only function of a filter for a sine wave is to filter out the fundamental, in other words, the note itself).

The filter then feeds into the Amp envelope which is set as follows:

- Slope Level: 100%

- Decay Time: 0.84 seconds

- Decay curve: exponential

- Sustain Level: 0%

- Release Time: 0.58 seconds

- Release curve: exponential

- Amount: 100%

This gives a fairly basic tom sound. While I liked this, I wanted something more and so I loaded a Vintage Square 1 wave into oscillator one and fed this into filter one. The filter does have some effect on a square wave. In this case, I set the filter as follows:

- Type: 12 dB low-pass

- Cut-Off: 605 Hz

- Resonance: 38 dB

This gives a different tone and texture when compared with the sound created by the sine wave. However, I wanted to limit this sound to the attack only, so I set Envelope one to control oscillator one's Level. I then set Envelope one as follows:

- Slope Level: 100%

- Decay Time: 0.01 seconds

- Decay curve: exponential

- Amount: 100%

I then set the tone of the tom by balancing the Levels of the two oscillators. With oscillator one's Level set to 100% and oscillator two's Level set to 75%, I got the sound I wanted. You can tweak the respective levels to adjust the tone further.

Chapter 12—Refill #4

Making Sounds with Wusikstation

As some of you may know, I wrote the original user manual for Wusikstation. As part of that manual, I included a sound design tutorial. That manual was first written when Wusikstation was still in beta. At the time of writing this book, Wusikstation version 4 is in beta, and so it is probably time to update that original tutorial.

For anyone who read the original, you will see that there are some similarities in this refill. However, given that this refill is attached to a whole book about synthesizer programming and we have already designed many Wusikstation patches through the book, I feel it is unnecessary to walk you through the first steps of sound design again. Instead, this tutorial will introduce you to the practical application of some of the techniques set out in the rest of this book.

If you haven't read the rest of the book, then you might find that this refill introduces some unfamiliar concepts. If you are looking for the other Wusikstation patches, in the hard copy book you can check out chapter 8, chapter 12, and refill#1.

Bass Sounds in Wusikstation

I will start this refill by creating some bass sounds. Before I do this, I want to think about what the bass part does. Conventionally in any track, the bass part will often perform one of two functions:

- First, it can act as a rhythm part. Usually you will find that the bass part works with the kick drum to provide the fundamental rhythm of a track. You know the fundamental rhythm: that's the thing that people dance to—it's the real groove of the track.

- Second, a bass part can act as part of a chord, to provide the harmony. In this role, the bass will most often work to provide a root note for any chord.

It may perform other functions, but, these are the main two, and so I want to focus on them.

Of course, it goes without saying that these two parts are not mutually exclusive: a bass part can provide rhythm and harmony. However, you will often make different decisions about how to program a bass sound depending on how the sound is going to be used. Let me explain further in the next two sections.

Bass Part Used for Rhythm

When a bass part is used for rhythm, it is often necessary to ensure it is heard as well as felt. To do this, the attack portion of the note needs to be audible. This is usually achieved by modulating a low-pass filter with an envelope. This modulation also allows the timbre of a note to change over time, much like an acoustic instrument.

It is important to get the attack right to ensure it fits with the track when the bass has a rhythmic role. One of the keys to getting the right sound is to control the decay time of the envelope that modulates the filter. This will control the apparent "length" of the bass note. If the decay time is too short, the note will be too staccato. If the decay time is too long, then the note will be flabby and lack definition.

If you're using your bass part as a rhythm part, then the groove is essential. Apart from the melody, you can pretty much forget everything else apart from the groove. If the groove doesn't work, then your track doesn't work. You may have the greatest fills and some fabulous hooks, but if you don't have the groove, your track won't have a pulse. No pulse = no life.

Now, before I talk about sound design, a plea on my part. Please remember why people dig music: it is because the music moves them somehow. People (except in a few cases) are not interested in how a sound is created—they just care about the groove and the melody.

So when you come to create a sound, remember, the actual sound is a lot less import than how well the part works in the context of your track. This is particularly so for a bass track where you can use a simple sound.

So when you come to program a bass sound, I suggest you get the bare bones of your track up and running—a bass part and a kick drum should be sufficient—and load up a very simple bass sound. For an example of a simple bass sound, check out 01 Rhythmic Bass Start Point.

This patch has a Sawtooth wave (from the Soundsets/Waveforms folder) going through a 24 dB/octave (4 pole) low-pass filter. The filter's Cut-Off has been set to zero and the Resonance has been set to a low level (in this case 20). On its own, this gives a really tedious sound and so to add interest, the filter is modulated by Mod Envelope 1. It is Mod Envelope 1's setting—in particular, the short Decay time (31), and the zero Sustain level—that give this sound its characteristic.

If you play the sound, you'll hear a very short note. Mod Envelope 1 is the only envelope that has any effect on this sound. The Amp (volume) Envelope for this oscillator has its Attack time set to zero and its Sustain level set to maximum, it therefore has no effect on the volume of the sound.

I've set the Mod Envelope 1 curve to Exponential ×1 to give it a faster and smoother decay time. In the modulation matrix, Mod Envelope 1 has been set to modulate Filter 1 Frequency (that is filter one's cut-off frequency) and the Amount of modulation has been set to the maximum (127).

So what do you do with sound? How does it help you?

Well, as I said before I distracted myself explaining how the sound was created, you should load this sound in a track and play just the bass and the kick drum part. Then tweak the bass sound to help it fit better in the track. For this exercise, you can tweak three settings:

- the Decay time in Mod Envelope 1

- the Sustain level in Mod Envelope 1, and

- the Amount (by which Mod Envelope 1 modulates Filter 1 Freq) in the modulation matrix.

Once you've tweaked, listen to your kick drum and bass sound. How does it sound? Is it good enough? Does it really hold down the groove?

If it doesn't, then don't blame the sound. There's nothing wrong with it: it's the groove that's at fault. Go back and work on your groove until it really does *groove*. Don't move on until you've got that groove. Until the track works because you have chosen the right (bass) notes, played at the right time, then it's not worth pursuing the sound. If a part only sounds right because you have picked the right preset (or programmed a great sound) then you have a fundamental problem with your track.

Once you've written a great part and tweaked the patch so that it fits, then your work is done: your bass sound is programmed. However, later in this refill I'll talk about a few things you can do to get a more interesting sound if that is needed.

Bass Part Used for Harmony

Now instead of using your bass note for rhythmic effect, you may use it as part of a chord. In particular, you may use it as the root note of a chord.

The role of determining a chord should not be overlooked. Play a simple triad of C, E, and G and you've got a C chord. Now add an A bass note and you've got an Am7 chord which has a very different characteristic.

For this sort of bass sound, the pitch needs to be clear and the bass note should not dominate: it should be part of a chord. If you've got the patches, take a listen to 02 Soft Bass Start Point and you will hear a bass sound that can be used for this purpose.

While it may not sound it, this patch is very similar to the previous one. All I did to create this was take the previous patch and tweak it. In particular, I tweaked Mod Envelope 1 to remove some of the attack from the sound (by setting the Attack to 33) and I also dropped the Amount in the modulation matrix by which the envelope modulates the filter to 35. One other more subtle tweak was to increase the Resonance to 37 to give the sound a touch more twang.

As you can hear, the result is quite a subdued brassy sound which could work as a start point for programming a harmony bass sound in a range of styles. When you are programming this type of sound, there are a few things to think about:

- First, before you program any sounds, consider whether the easiest solution is to use a lower note in an existing part as your bass note. For instance if you have a keyboard part already, could you just add a bass note here. The advantage of this approach is speed and simplicity. The disadvantage is that you create a homogenized sound.

- Make your sounds as simple as possible. I will discuss some tweaks to make your sounds more interesting. However, remember to always program your sound while the track is playing (in other words, as your listener will hear it) in order to get the best sound. If you program your sound with no other instrument playing, you will be tempted to make the sound more "interesting". If you need more "interest", then rewrite the part, don't rewrite the sound.

- While you are rewriting your part, remember that simple, less busy bass parts often work best. The point of the bass part is to support the track, not to be a feature on its own.

- Make sure the attack time and the release time are not too fast. If these are too sharp, the bass sound can have too much rhythmic effect.

Variations on a Theme

So now you understand what your bass part is doing and the groove of your track is positively smoking. You've got a bass sound that works within the track and you've created a bass sound without working too hard.

However, once you've got your track up and running, you're likely to want to add some variation. Perhaps you might want to change the bass sound in the second verse to give a bit of variety or maybe you feel that the bass part doesn't fill up the track sufficiently in your club mix when everything else apart from the drums drops out for that "hands-in-the-air" moment.

The temptation will be to use another bass sound and perhaps to create a second bass part with a new bass sound. However, can I suggest something you try different? Instead, look to your existing sound and automate it. In other words, change your existing sound. The most obvious things to change will be the filter's cut-off frequency and resonance.

These are not the only features of a sound you can change. As I mentioned earlier on, a key factor in determining the character of a sound is the decay time setting. Just for fun, take a track and add the sound 03 Tweak the Decay. This is the 01 Rhythmic Bass Start Point sound that we used earlier on but I've tweaked the Mod Envelope 1 Decay time, Sustain level, modulation matrix Amount, and the filter Resonance amount to give a sound that works for me. If you don't like it, please feel free to tweak the sound some more until it works for you.

Now take the sound, load it into a track, and automate the Mod Envelope 1 Decay time as the track plays. So for instance you could make the decay time shorter in the verse to create more tension through a more staccato sound and then increase the Decay time (just slightly—you don't need dramatic changes here) in the chorus to give a fuller, fatter tone.

Play with it and see what works for you.

Where Will the Sound Be Listened To?

When you're creating and tweaking your sound, take some time to think about how your track is going to be listened to. For instance, if the track is going to be heard through an iPod/mp3 player headphones, then there is little point in worrying about the low end of your sound since the headphones will not be able to cope with the low frequencies.

Alternatively, your track may be destined to be heard in a club. This brings a whole new bunch of challenges. Perhaps the biggest challenge is how much is too much bass? Here I'm going to duck out of offering any advice and suggest this is an area that comes with experience.

There are other challenges with creating sounds for a club. One issue is the stereo separation. Often in a club it's difficult to get any real stereo separation due to the positioning of the speakers and so you will probably want to put your bass sounds towards the middle of the stereo spectrum so that people don't lose the bass end if they can only hear one side of the stereo spectrum.

Also, if your track is going to end up on vinyl, then you need to ensure the bass end is in the middle of the stereo spectrum. Vinyl is a physical medium. If you have an unbalanced bass sound—for instance, if you have two slightly detuned oscillators pushed to the left and right channels respectively—then the phase difference through detuning may make the needle jump. To counter this, if you're creating a sound that will be played on vinyl, you might want to keep your bass sounds panned to the centre.

For anyone old enough to remember the days before CDs, this suggestion will not be news. It is only since digital music has become widespread that mixing engineers have considered moving the bass part away from the centre of the stereo spectrum, and indeed, many people still consider that the bass and kick drum should stay in the centre of the spectrum to spread the weight over the speakers.

Thickening Up the Low End

So by now you will have sorted out your bass part and fixed your mix. What if the bass sound you have created still isn't right? What happens if it still isn't richer, deeper, thicker, brighter, or sharper?

Well, then it's time to get tweaking again. However, when you get tweaking, remember that the bass sound must still fit in your mix. It's pointless creating a "feature" bass sound if the rest of your track sucks.

Now we all know that the easiest way to thicken up a sound is to double it and to slightly detune the second oscillator, perhaps panning the two oscillators in opposite directions in the stereo field. However, as I have already noted, this may not always be the best course so here are some other ideas you can use to add weight to your sound.

High-Pass Filter

It seems totally counter-intuitive to use a high-pass filter with a bass sound, but this is a very effective technique for boosting a sound without changing its character.

Take a moment to think about a filter—the resonance control allows you to boost the sound around the cut-off frequency. With a low-pass filter, sounds above the (boosted) cut-off frequency are then reduced. A high-pass filter works in reverse by cutting the sound below the cut-off frequency.

But wait a moment, what about the resonance control for a high-pass filter? This is where the magic comes in!! By boosting the resonance and setting the cut-off frequency to track the pitch of the note, the high-pass filter boosts the bass. With Wusikstation, the pitch tracking can be set up through the modulation matrix with the source being pitch and the destination being the filter.

In a bass sound, we care about the fundamental frequency, and we don't usually care too much about any frequencies below that, so a high-pass filter is ideal. You may be wondering why we didn't just use EQ here to boost the low end. The key here is the key tracking—EQ doesn't key track. A high-pass filter can key track and so it is only the root note of the sound that gets boosted. EQ could give a wooly low end and also won't boost the fundamental of the note at higher pitches.

If you load up 04 High Pass Thickening you can hear this effect in practice. This patch takes 03 Tweak the Decay and adds a high-pass filter (in filter 2) to boost the bass. In this case, the high-pass filter is a 4 pole filter with the Cut-Off set to zero and the Resonance set to 60. Switch between the two patches and you'll hear quite a pronounced effect. As you can see this sound uses two filters: the low-pass filter shapes the sound and the high-pass filter adds weight.

Sine Wave

A sine wave on its own may be quite dull and may not seem to be an appropriate wave for a bass sound, but when you add it to an existing patch, it can add some floor-shaking sub-sonics. A sine wave used in this way has the added advantage of not changing the tone of a sound too much.

Listen to 05 Plus Sine and you will hear an example of a sine wave working to emphasize the fundamental frequency and so thicken a sound. For this patch I took 03 Tweak the Decay and added a pure sine wave in oscillator 2. The filter in oscillator two is switched off and there is a small amount of volume shaping with that oscillator's amp envelope.

Sub-Oscillator

You aren't limited to a sine wave if you want to add weight: you can also add a sub-oscillator. A sub-oscillator is a simple oscillator usually added one or two octaves below the main oscillator.

Load up 06 Plus Square for an example of a sub-oscillator patch. Here I have taken 03 Tweak the Decay and added a square wave (using the Pulse waveform from the Waveforms folder) an octave below the main oscillator. This square wave then passes through a low-pass filter which has its cut-off frequency modulated by Mod Envelope 1 much in the way that oscillator 1's filter is modulated.

The effect of adding the square wave is to thicken up the sound. However, there is a side-effect in that the tonal characteristics of the sound are changed. This may or may not be a problem. If you like the sound and it still fits well in the mix, you don't need to worry. If you don't like the sound, or you are having trouble getting it to sit in a mix—perhaps the sound with the square wave might get too dominant—then you will have to look at one of the other options for thickening up your sound.

While I have used a square wave here, you are not limited to that waveform as a sub-oscillator, for instance, sawtooth waves work well, too. Take some time to listen to the other waves and you will find which works for you and which don't.

Adding More Presence

We've looked at how to add weight to your sound. So what happens if your bass can be felt, but can't be heard?

Clearly you've got several choices, for instance:

- Open up the filter a bit to make your sound brighter. You can do this by tweaking the cut-off frequency, or adjusting the filter's modulation by an envelope (or any other modulation source).

- Double your oscillator, but instead of slightly detuning the second, raise its pitch by an octave.

Either of these choices might work. However, they will both change the tone of your sound which will mean the patch dominates the sound spectrum to a greater extent. As with the thickening options, this may be a good or a bad thing depending on the sound you are looking for.

One other option is to add some distortion. This might sound quite radical, but it is actually a very controllable effect. Check out 07 Distortion to hear this effect in practice.

For this sound I have taken 03 Tweak the Decay and have added the Digital Distortion effect in FX1. The amount of distortion is controlled by the FX1 Send control. For this patch I have included quite a bit of distortion so that the effect is noticeable and the tonal characteristics of the sound are changed. If you play with the FX1 Send control you will be able to hear that at quite low levels there is a brightness added to the sound without too much distortion. Alternatively you might like the higher amounts of distortion which really help the bass sound to cut through!!

Pad Sounds in Wusikstation

That's enough bass. Let's create some pads. As with the bass sounds, when creating pads, you need to exercise some caution in order to ensure that your sounds don't dominate the whole track (unless that is your intention).

08 Release me

There are two parts to this patch:

- a fairly plain "pad" type sound (perhaps reminiscent of synthesised strings), and

- some bell like fairy dust.

The interaction between the two elements is what makes this pad interesting. The pad element is fairly constant. It is based on two sawtooth waves (Saw PW in the Waveforms folder) in oscillators two and three. The second wave has its pulse-width modulated—in the modulation matrix, LFO 1 modulates layer two's loop position: this is what achieves the PWM effect. This element fades in (layers two and three have slower attack times) and is slightly velocity sensitive.

The fairy dust element is based on the Arps16 wave (in the Famous Keys Volume 4 bank). No clever effects are used to get the sound, it is the sample alone that creates that sound, however, Modulation Envelope One does some interesting things:

- First it acts to suppress the volume of the wave. In the modulation matrix the Amount setting is at -127.

- Second, the envelope is velocity sensitive: this means that the amount of suppression is controlled by the velocity. With light playing (low velocities) the wave is not suppressed much and can be heard, with high velocities the wave is suppressed completely.

- Third, there is a long decay time—this means that over time the suppression fades and the wave can be increasingly heard (the Sustain level is set to zero—when this is reached there is no suppression).

- Fourth, the release time of this modulation envelope is much faster than the release time of the amplifier's envelope. This means that when the note is released the fairy dust sparkles will be heard until the release portion of the amplifier envelope has ended.

This patch has been designed to be played—you will get the best performance from it using a velocity sensitive keyboard. Contrast short, staccato stabs with some longer held chords.

You will also hear a considerable amount of echo has been added to this patch. This sounds great when the patch is heard on its own. However, if you were to use a patch of this nature in a mix, you may want to lower the echo or filter some of its low frequency content.

09 Slow & warm

This is another pad that has been designed with the player in mind. As the patch name suggests, it is a slow, warm sounding pad based on two Super Saw waves (from the Famous Keys Volume 2 bank) which have had most of their top end filtered out using a 2 pole filter with the Cut-Off set to 19 and 22, respectively.

There are two particular features added for the player:

- first velocity opens up the filters—play harder for a brighter sound, and

- second, the modulation wheel closes the filters—push up the modulation wheel for a duller sound.

To give the patch some movement:

- a Modulation LFO has been applied to the pitch of layer one, and

- another modulation LFO has been applied to the volume of layer two.

As with the previous patch, a fair dollop of echo has been added that you might want to turn down if you were to use a similar sound patch in a mix.

Zoning in Wusikstation

10 12 Pluck demonstrates the zoning feature of Wusikstation to mimic the behavior of a 12 string guitar.

10 12 Pluck

There are three layers in this patch, all based on the WS01 wave in the Famous Keys Volume 3 folder:

- Layer one plays across the whole of the allocated range of this patch (note 40 to note 112).

- Layer two, sounds an octave higher than layer one and is allocated the range from note 40 to note 82.

- Layer three is the same pitch as layer one (give or take a bit of detuning) but only sounds over the range from note 83 to 112.

Layers two and three are mutually exclusive—they have been set up in a way to mimic the doubled strings on a 12 string guitar.

Frequency Modulation Synthesis in Wusikstation

I have looked at FM techniques in several places in this book. To end this refill, I now want to focus on FM and Wusikstation. My aim here is to demonstrate how to apply FM techniques in Wusikstation.

Now, before we go any further, I should caution you against expecting to be able to recreate all of those wonderful FM sounds you have heard from the old 1980s records with Wusikstation. Wusikstation is a great synth, but it would not be my first choice synthesizer for creating those classic/cliché FM sounds. However, it does create a wide range of sounds that have their own unique character which is why I believe Wusikstation is useful as an FM synth, and so for the remainder of this refill, I am going to focus on the range of FM tones that Wusikstation can readily create.

If you want to read more about FM theory, then go back and check out Chapter 7: Frequency Modulation Synthesis.

I'm now going to create some very basic FM sounds. The purpose of these sounds is to show you how you can create FM sounds in Wusikstation and to introduce you to some of the tones that are available.

The key range over which these patches can work is variable—that is in the nature of FM sounds. Accordingly, I have restricted some of the key ranges of these patches. If you can't hear anything, then try a different (lower) key. You will also see that I have not introduced any key scaling. If you're of an ambitious nature, set these patches with a full key range and tweak the key scaling parameters!!

11 Simple FM

11 Simple FM is our first FM patch in Wusikstation. Oscillator two works as the modulator and oscillator one is the carrier. Both oscillators are pitched at the same level, and the volume of oscillator two has been dropped to zero so all you hear is the FM sound.

The FM sound is created through the modulation matrix:

- The modulation Source is O2 Out.

- The modulation Destination is O1 Pitch.

- Min and Max are set to 0 and 127, respectively.

- Amount, in other words the amount by which oscillator two frequency modulates oscillator one, is set to 62.

In this patch, the FM Index (the amount by which the modulator modulates the carrier) is fixed by the Amount setting. You will find that you can adjust the tone of this patch by adjusting this value.

12 FM Vel to Mod

In this second patch, 12 FM Vel to Mod, the amount by which the modulator modulates the carrier (the FM Index) is controlled by MIDI velocity, so:

- at lower velocities, the FM Index is lower, and the tone is duller, and

- at higher velocities, the FM Index is higher, and the tone is brighter.

This patch is exactly the same as 11 Simple FM, but with one tweak in the modulation matrix to allow velocity to control the FM Index. To achieve this, I added another line with the following settings:

- The modulation Source is Velocity.

- The modulation Destination is O1 Pitch.

- Min and Max are set to 0 and 127, respectively.

- Amount is set to 62.

As with the earlier patch, you can control the maximum extent to which velocity can control the sound by tinkering with the Amounts (in either, or both, lines).

This patch also highlights another one of those "challenges" when dealing with FM: pitching. As you play this patch with varying velocities, you will hear that the pitch appears to change. While this is a side effect of the FM technique, the phenomenon may be more noticeable in Wusikstation that is the case for Rhino, Surge, or Z3TA+.

13 FM Env to Mod

In this third patch, the amount of modulation (the FM Index) is controlled by an envelope.

Introducing an envelope in this manner allows for the tone to shift over time. In this case, I have selected an envelope shape to give an initial impact and a rapid decay. This effect is created by using the Amp Envelope for the modulator (oscillator two).

This patch is very similar to 11 Simple FM, the only difference is in the modulation matrix where only one line is necessary with the following settings:

- The modulation Source is O2 × Env, so the output of the modulator is controlled by its envelope. As there is an amount of velocity scaling in the envelope, this translates through into the FM Index too.

- The modulation Destination is O1 Pitch.

- The Amount is set to 127.

For this patch the FM Index is most immediately controlled by the Amount setting, and also the Velocity setting in the modulator's envelope. However, the Sustain level of the envelope also has an effect on the sound, determining the tone during the sustain period of the envelope.

You will also see that I have used the Exponential ×2 curve in the envelope to give a smoother decay.

Varying Modulator : Carrier Ratios

In the patches we have looked at so far, the carrier and the modulator have had the same pitch. Any tonal variation has been controlled by the FM Index (that is, the modulation depth) which has been controlled by the Amount setting in the modulation matrix, and influenced by velocity and an envelope.

I now want to look at a number of patches where the pitch of the modulator and carrier are different. For each of these patches, the pitch of the modulator and carrier are fixed, however, the amount of modulation is controlled by velocity.

Listen to:

- 14 Mod -24 where the modulator is pitched two octaves below the carrier

- 15 Mod -19 where the modulator is pitched one octave and seven semitones below the carrier

- 16 Mod -12 where the modulator is pitched one octave below the carrier

- 17 Mod +7 where the modulator is pitched seven semitones above the carrier

- 18 Mod +12 where the modulator is pitched an octave above the carrier, and

- 19 Mod +19 where the modulator is pitched one octave and seven semitones above the carrier.

In very broad terms, you can hear that when the modulator is pitched below the carrier, there is a darker tone, but when the modulator is pitched above the carrier, there is a brighter, sometimes metallic, tone. Also, as the modulator is pitched higher, the sound becomes less controllable (hence the reduced key ranges for some of the later patches).

And Finally...

Further Reading

If you have enjoyed this book and found it useful, you might want to check out some of my other books:

Cakewalk Synthesizers: From Presets to Power User

ISBN-10: 1-59863-314-7, ISBN-13: 978-1-59863-314-6

Cakewalk synthesizers are some of the most powerful and most widely used software synthesizers on the market. This book shows users how to operate and get the best results from Cakewalk's complete range of synthesizers (including those that come with SONAR, Project5, and Kinetic, the rgc:audio synthesizers such as Z3TA+, and Cakewalk's stand-alone software synthesizers such as Dimension Pro and Rapture).

This is the ultimate guide to understanding and using all of Cakewalk's synthesizers and learning about synthesizer programming. For further details, go to www.noisesculpture.com/cakewalk

Building a Successful 21st Century Music Career

ISBN-10: 1598633708, ISBN-13: 978-1598633702

Building a Successful 21st Century Music Career is about how you can proactively start, and manage your own music career so that you can earn your living through making music—whether as a musician playing pop, rock, soul, R&B, classical or any other style of music; as a songwriter or composer; or as a producer.

For further details, go to www.noisesculpture.com/c21

My Website

You can always find further details about all of my books at www.noisesculpture.com

Appendix
The Synthesizers

The examples in this book have been based on six software synthesizers.

The synthesizers I have chosen are some of the best that are currently available. They have all been built by talented, passionate developers who have designed something unique and brilliant. These developers are all musicians who have built a synthesizer that they want to see built.

This book is not a review. You will see that I have highlighted many of the positive aspects about all of the synthesizers: this book is about how to do things—specific aspects of the synths in question have been used to illustrate my specific points. It is not a review and my intention is not to highlight any features that you may perceive as a weakness.

I am sure that if you talk to the developers they will tell you that I haven't highlighted some really cool features of their synths and they would be correct in this assertion—this is not a book about synths: it is a book about sound design. If you are looking for book about how not to do things, the limitations of software synthesizers, and what you wish software would do for you, then this is the wrong book for you.

If you don't already own all of the synths, I suggest you do. If nothing else, get hold of the demos and find out how each synth really works. The patches will work with most of the demos (although for Wusikstation some of the waveforms used in the patches are only available in the purchased version, and Rhino requires the full version to be purchased before external patches can be loaded).

The summary below only lists the main features of each of the synths. The prices listed are those quoted on website at time of writing (June 2007)—note, some prices are quoted in US$ and others are quoted in EUR€.

Cameleon 5000

Developer: Camel Audio

Website: www.camelaudio.com

Price: $199

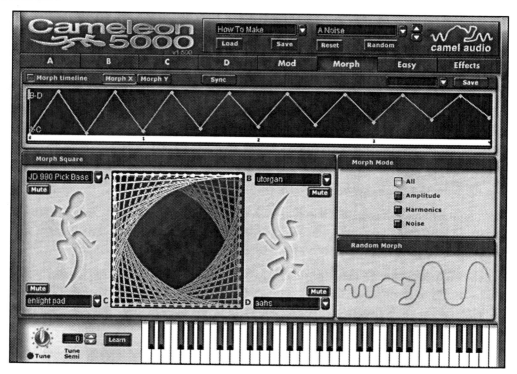

Cameleon 5000 was one of the first software based additive synthesizers and, I believe, was the first additive synthesizer to offer practical resynthesis.

At its heart is the morph square which allows users to morph between four sounds at once (including user imported sounds). Once imported, sounds can then be edited in ways which are impossible with a conventional sampler, for instance there is control over each of the harmonics. As you would expect there are the usual LFOs, envelope controls and FX units.

Definitely one for the creative and the curious.

Rhino

Developer: Big Tick

Website: http://bigtick.pastnotecut.org

Price: €70

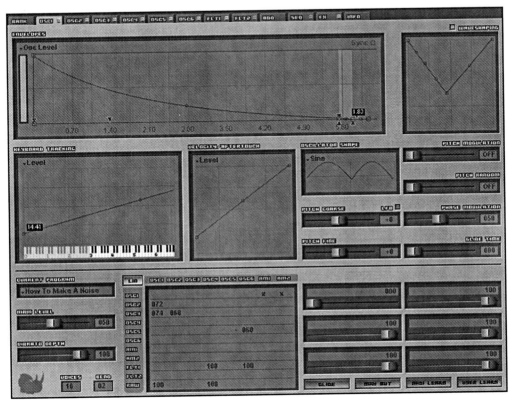

Rhino comes with six oscillators (offering more than 130 built-in waveforms and the ability to import your own) and a built in additive waveform generator. It has two filters and countless graphical multipoint envelopes with curvature controls. Add in a step sequencer and one variable wave shaper per oscillator and a visible output routing matrix and you have a hugely powerful machine.

However, even with all this power, for me, Rhino is still the ideal FM machine.

Surge

Developer: Vember Audio

Website: www.vemberaudio.se

Price: $150/€135

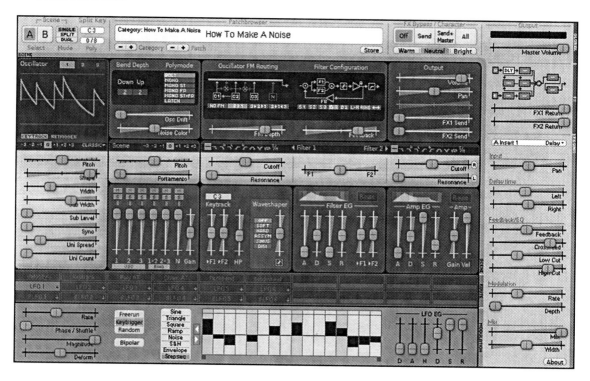

Surge is the new kid on the block, but is a hugely powerful synthesizer with six oscillators offering countless numbers of waveforms. It has leading edge features, great performance, and a highly intuitive interface making it a highly desirable synth.

Vanguard

Developer: reFX

Website: www.refx.net

Price: $89.99

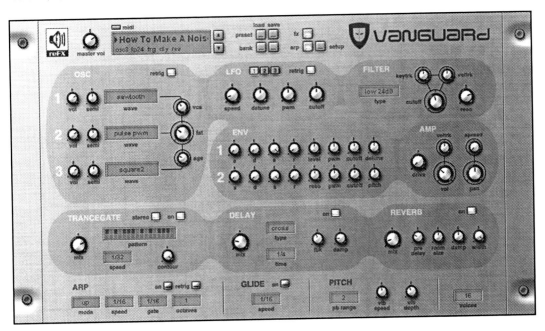

At first look, Vanguard appears to be a simple subtractive synthesizer and it is certainly capable of producing warm, rich sounds. However, it comes with 31 different oscillators and a selection of 13 filters (including some combined dual filter variations) giving it a wide range of sounds.

A key advantage of Vanguard is that you can see what's going on—virtually all of the controls are accessible at the same time. The only controls that aren't immediately accessible are the second and third LFOs.

Vanguard is incredibly easy and fast to program and creates great sounds: for me that often makes it a first call synthesizer.

Wusikstation

Developer: Wusik dot com

Website: www.wusik.com

Price: $99.95 (although there are often special discount offers giving a lower price)

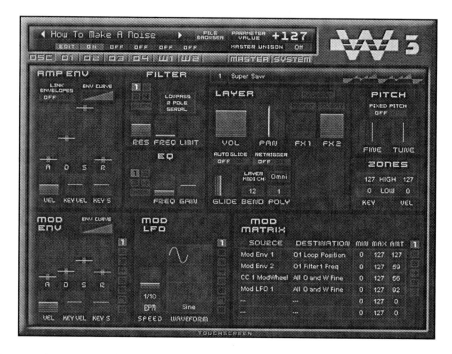

Wusikstation has six sample based layers. Each layer can load separate multi-sampled soundset which is can then be processed through up to four filters. As it is sample based, the waveform options are effectively limitless. Each layer has its own envelope and there are also eight modulation envelopes and eight modulation LFOs which can control a variety of destinations through the modulation matrix.

The two wave-sequence layers offer possibilities for the creation of unique and complex sounds. Its low CPU usage and ability to create compelling sounds makes Wusikstation a versatile synthesizer which can be used in many musical situations.

Z3TA+

Developer: Cakewalk

Website: www.cakewalk.com

Price: $99.00

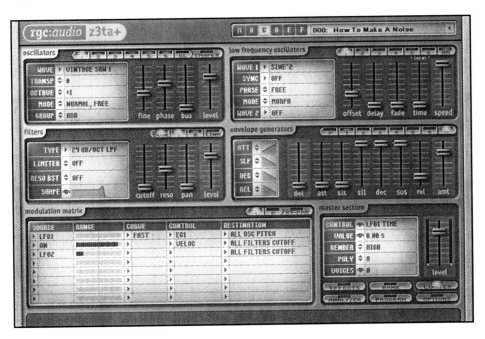

You could pretty much fill a book just talking about the features of Z3TA+. Behind the unassuming GUI is a real heavyweight synthesizer.

So what does Z3TA+ offer? Full stereo processing. 6 oscillators, 60 built-in waveforms, 6 user waveforms, PWM possible for all oscillators with any waveform. Independent waveshaper for each oscillator with 14 wave transformations. Multi-mode for any oscillator. Individual settings for pitch bend up and down (-12 to +12 semitones). 2 Stereo filters. 6 morphing capable LFOs. Eight 6-stage envelope generators. Arpeggiator/sequence player. Modulation matrix. Many effects. 768 program capacity (in six banks). To read the full list of features, check out the website.

There are two main reasons to use Z3TA+. First, and most obviously, the sound: this is perhaps the benchmark against which all other software synthesizers are judged. Second, it can do pretty much anything. Its depth (in sound and features) and usability for musicians make it a synthesizer that should be in every musician's arsenal.

Printed in the United Kingdom
by Lightning Source UK Ltd.
130104UK00001B/224/A